PALAEONTOLOGICAL ASSOCIATION
FIELD GUIDES TO FOSSILS: Number 8

The Jurassic Flora of Yorkshire

By

J. H. A. VAN KONIJNENBURG-VAN CITTERT
Laboratory of Palaeobotany and Palynology,
University of Utrecht,
Budapestlaan 4, De Uithof,
3584 CD Utrecht, The Netherlands

and

H. S. MORGANS
Department of Earth Sciences,
University of Oxford,
Parks Road, Oxford OX1 3PR, UK

THE PALAEONTOLOGICAL ASSOCIATION
LONDON
1999

ISBN 0-901702-64-1

Editor C. J. Cleal
*Department of Biodiversity and Systematic Biology,
National Museums and Galleries of Wales,
Cardiff CF1 3NP*

Series Editor David K. Loydell
*School of Earth,
Environmental and Physical Sciences,
University of Portsmouth,
Burnaby Building, Burnaby Road,
Portsmouth PO1 3QL*

Printed in Great Britain by Henry Ling Ltd, at the Dorset Press, Dorchester, Dorset

CONTENTS

ACKNOWLEDGEMENTS

The authors thank Dr J. T. van Konijnenburg and Mr D. Smit for their help with the drawings, and Mr H. A. Elsendoorn for his work in the darkroom. Part of this guide is based on research by H. S. Morgans, funded by the N.E.R.C., and carried out at the Department of Earth Sciences, University of Oxford. Dr S. P. Hesselbo and Prof. R. A. Spicer are thanked for their guidance during this work. This publication is NSG (Netherlands School of Geology) paper no. 961107.

1. INTRODUCTION

THE Middle Jurassic Ravenscar Group of Yorkshire yields some of the most abundant and diverse plant fossils found anywhere in Britain. Some 260 species have been collected from more than 600 plant beds in the region, a number which reflects the flora's long and detailed history of research. The investigation of the flora can be traced back to the studies of Young and Bird in the early part of last century (1822), and features numerous subsequent inquiries. Of these studies, the work of Tom Harris is perhaps most notable, well illustrated and documented in five volumes entitled *The Yorkshire Jurassic flora* (Harris 1961, 1964, 1969, 1979; Harris *et al.* 1974). Despite such exhaustive study, the stratigraphical context of the flora (and main plant beds) has until now been surprisingly poorly understood. Thus, although the nature of the Mid Jurassic vegetation is fairly well constrained, spatial and temporal variations in the plant assemblages are not. One of the main objectives of this guide is, therefore, to provide stratigraphical constraints, and a sedimentological background, to four of the best known plant beds in Yorkshire (at Hasty Bank, Hayburn Wyke, Gristhorpe and Scalby Ness).

Inspiration for this guide stemmed from a visit to the Yorkshire flora by the Linnean Society Palaeobotanical Specialist Group in April of 1994. For this trip a brief guide was compiled by the authors, upon which this present work greatly elaborates. In the following chapters detailed taxonomic lists and key descriptions of the more common plant fossil species are provided within the context of the four chosen plant beds. The richness and diversity of the vegetation growing *c.* 165–180 Ma (Gradstein *et al.* 1995) in Yorkshire is well recorded by these beds, which span the stratigraphical extent of the Ravenscar Group (Text-fig. 2). It is hoped that this guide will prove helpful in the preliminary identification of plant fossils collected from these localities, and offer useful information as to their environment of deposition.

With the exception of Hasty Bank, which lies inland, all localities are coastal, chosen not only for their historical interest but also for their good exposure and relative ease of access. However, details of routes to localities given by this guide do not imply a right of way. Users of this guide should therefore observe public rights of way detailed on the appropriate Ordnance Survey maps (sheet 93 Middlesbrough and Darlington; sheet 94 Whitby; sheet 101 Scarborough). To examine Hasty Bank, permission must be sought from the head forester at the address given in the relevant section below. Before visiting the coastal localities local tide tables should be consulted and, unless there are escape routes close by, it is advised that work is not begun on a rising tide.

2. GEOLOGICAL BACKGROUND

Stratigraphy

The Ravenscar Group consists largely of fluviatile and deltaic sediments, laid down during the Mid Jurassic in a gently subsiding region known as the Cleveland Basin (and today embodied by Yorkshire; Text-fig. 1). It is within the non-marine parts of this 250 m thick succession, referred to the Saltwick, Cloughton and Scalby formations (Text-fig. 2), that the most abundant and best preserved plant fossils occur, as finely preserved vegetative fragments, inflorescences, fructifications, spores and pollen. The rather sporadic presence of such well preserved plant remains contrasts with the ubiquitous occurrence of comminuted plant debris apparent in most parts of the group, including the marine beds. The Eller Beck and Scarborough formations represent the thin, although laterally extensive, marine intercalations within the group, deposited during

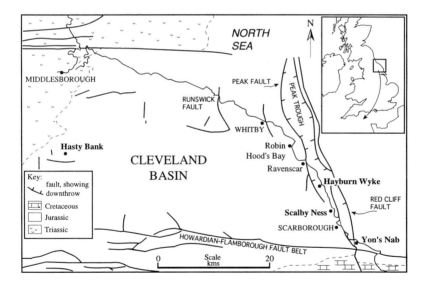

TEXT-FIG. 1. Map of North Yorkshire showing some of the places mentioned in the text; those highlighted in bold are the four plant-bearing localities discussed in this guide (map is modified from Rawson and Wright 1992). Text-figures 4, 7, 8 and 10 show the localities in more detail. The simplified structural framework is after Kirby and Swallow (1987) and Milsom and Rawson (1989).

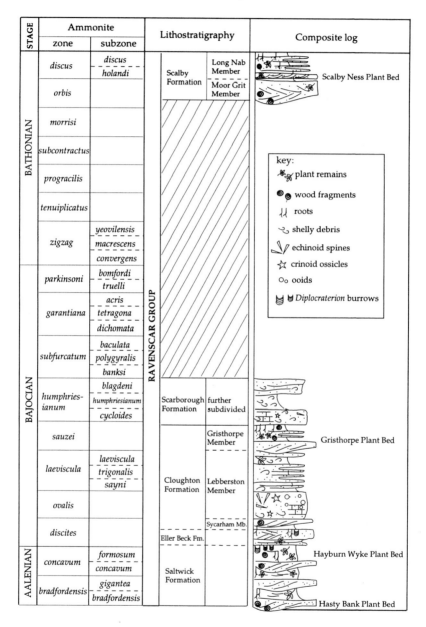

TEXT-FIG. 2. Simplified stratigraphy of the Ravenscar Group, showing position of the plant beds discussed in this guide.

intermittent marine incursions into the basin from the south and east. An additional, more laterally variable marine sequence is interbedded within the Cloughton Formation and known as the Lebberston Member (Text-fig. 2). Current stratigraphical practice (Hemingway and Knox 1973) employs distinctive sedimentary facies (lithostratigraphy) and characteristic fossil content (biostratigraphy) in order to subdivide the succession into separate stratigraphical units.

Despite the great variety of plant macrofossils contained within the group, they do not impart useful biostratigraphical information. Instead, the sequence is dated primarily using rare ammonites and ostracodes (e.g. Bate 1965, 1967; Parsons 1980). Arguably, ammonites yield the most precise biostratigraphical information, giving the potential for accurate correlation with sedimentary sections elsewhere. Within the Ravenscar Group, ammonites occur only in the Scarborough Formation (Parsons 1977, 1980). Dating of the group as a whole relies upon ammonite finds from the underlying Dogger Formation (Kent 1980; Parsons 1980) and the overlying Cornbrash Limestone Formation (Wright 1977). Text-figure 2 illustrates the biostratigraphical scheme which constrains the Ravenscar Group.

The following pages present a brief introduction to the structural development of the Cleveland Basin and the depositional history of the Ravenscar Group. The section is intended simply as an overview, which is elaborated upon in the locality descriptions of the four chosen plant beds below.

Structural development
The history of the Cleveland Basin began towards the end of Permian with differential subsidence, which continued during much of the Jurassic and Cretaceous, until inversion halted the process in the late Cretaceous or Tertiary. The effects of subsidence during the Permian were, however, minor compared with those which caused the development of the basin during the late Triassic (Kent 1980). Deposition during this period was controlled to some extent by the structure of the buried Carboniferous landscape beneath (Kent 1980). The basin is bounded to the west and north-east by less subsident or uplifted blocks which formed the Pennine and Mid-North Sea highs. The southern limit of the basin is defined by the Market Weighton High, which during the early to mid Jurassic acted as a hinge zone in the area of eastern England and greatly influenced stratal thicknesses. The block is currently thought to be underlain by granite (Sellwood and Jenkyns 1975; Bott *et al.* 1978; Kent 1980; Donato 1993). To the south-east, the Cleveland Basin extends into the Sole Pit Trough of the Southern North Sea Basin (Kent 1980; Hemingway and Riddler 1982).

Few faults are known within the basin, except along the coast where a number of north–south oriented normal faults exist. The most significant are the Runswick, Whitby, Peak, Scarborough and Red Cliff faults which developed intermittently from the Triassic to the Tertiary (Milsom and Rawson 1989). Recent acquisition of seismic data has allowed the Peak Fault to be mapped offshore, revealing that the fault forms the western margin of the Peak Trough (Milsom and Rawson 1989). The trough is a 5 km wide, graben-like feature which is bounded to the east by a series of faults between Scarborough and Cayton Bay. East–west oriented faults in the Vale of Pickering were active in the late Jurassic and Cretaceous (Kirby and Swallow 1987). Subsidence was terminated by a phase of inversion either during the late Cretaceous and early Tertiary (Kent 1980), or during the Tertiary alone (Hemingway and Riddler 1982). This inversion resulted in broad uplift of the basin along a east-north-east–west-south-west oriented plane (the Cleveland axis) flanked to either side by minor domes and basins, such as the Robin Hood's Bay Dome (Rawson and Wright 1992). Several normal faults were also reactivated at this time, with reverse, strike-slip and oblique-slip motion.

Today, the Jurassic rocks exposed in the Cleveland Basin dominate the coastline from Teesside in the north to Filey in the south, and crop out westwards to the Hambleton Hills. To the south-west, the outcrop is reduced to a narrow strip which continues south via Market Weighton and South Cave into Lincolnshire.

Depositional history

The Saltwick Formation represents the initial progradation of the delta southwards and eastwards across the Cleveland Basin, forming the lowermost portion of the Ravenscar Group. The 57 m thick unit is best exposed between Ravenscar (NZ 980 016) and Hayburn Wyke (TA 013 969) on the coast, although scattered inland exposures also exist. Although fine-grained sediments laid down in various overbank environments form the greater proportion of the sequence, significant sandstone bodies also occur, reflecting deposition in distributary channels which dissected the delta-plain. Numerous channels exist at the base of the formation, many of which cut down into the underlying Dogger Formation. Within this portion of the formation at Hasty Bank (NZ 567 036) abundant plant remains are contained within flood–plain sediments adjacent to a channel body. The character of this plant accumulation is described in more detail below.

The sediments resting upon these basal deposits display a vertical transition in channel type, from major bedload channels found at the base, to fewer, more isolated, mixed-load channels towards the top. Similarly, a vertical transition is also apparent in the overbank sediments, from well-drained flood-plain sediments to more poorly drained marsh deposits. In the upper part of the Saltwick Formation there are fewer roots but an increased amount of drifted plant material, less evidence for desiccation and more soft-sediment deformation (Livera and Leeder 1981). Well preserved accumulations of drifted plant material can be found at Hayburn Wyke, one of the four localities detailed in this guide. Vertical changes in the character of the Saltwick Formation are interpreted to reflect the gradual abandonment of the delta lobe, which culminated in a rapid incursion of sea water across the delta-plain. This sea-level rise, originating in the east of the basin, resulted in the deposition of the 4–8 m thick Eller Beck Formation (Knox 1973; Hemingway 1974; Powell and Rathbone 1983). This formation is entirely marine and is composed of thin, intercalated ironstone and siltstone beds at the base, capped by wave-generated beds of sandstone displaying intense bioturbation and roots at the very top. This uppermost part of the formation is interpreted as a sandy barrier-bar or shoreline, which passed over the shallow-marine muds forming the lower part of the formation (Knox 1973). Progressive progradation of the delta resulted in vegetation colonizing the barrier-bar, evinced by the roots that we see penetrating the top of the sandstone today. These features are particularly well displayed at Iron Scar (TA 015 968).

The 50 m thick Sycarham Member, which rests upon the Eller Beck Formation comprises a succession of cross-bedded channel sandstones intercalated with mudstones and siltstones. It is best exposed (although still rather incompletely) on the coast between Iron Scar and Rodger Trod (TA 021 955). Fragmentary plant remains are common and are scattered throughout the sequence, but larger, better preserved plant fossils are comparatively rare. The overlying, laterally variable Lebberston Member constitutes a marine sequence within the middle part of the Cloughton Formation, and is separated into the Millepore Bed and Yons Nab Beds (Sylvester-Bradley 1953; Bate 1959). As a whole, the member is composed of a richly fossiliferous, sandy calcareous oolite laid down in shallow-marine to coastal settings in the south of the basin. It is correlative with thinner, poorly fossiliferous, bioturbated sandstones in the north, which were deposited in beach and lagoonal environments (Livera and Leeder 1981). Plant material in the Millepore Bed is sparse and more commonly found in the Yons Nab Beds, where it is scattered and fragmentary.

The succeeding Gristhorpe Member marks a return to deposition on the delta-plain. The sequence attains a maximum thickness of some 30 m and typically comprises a succession of crevasse-splay sandstones, intercalated with plant-bearing mudstones and siltstones laid down in a flood-plain environment. Evidence for channel development is rare, with the exception of a laterally accreted channel sandstone cutting down into the Yons Nab Beds from the base of the Gristhorpe Member at Yon's Nab (TA 081 844). Intermittent marine influences during the deposition of the member are indicated by the presence of marine trace fossil assemblages at certain stratigraphical levels. Despite this marine influence, the plant fossils found within the unit are rich and diverse, occurring most abundantly near the base of the sequence within the Gristhorpe Plant Bed.

The Scarborough Formation, which overlies the Gristhorpe Member, is the thickest marine unit in the Ravenscar Group, the environments of deposition including brackish sandy embayment, nearshore sandy and muddy shelf, and offshore mud-dominated shelf (Gowland and Riding 1991). Plant fossils are rarely found within the Scarborough Formation, but where they do exist they are small and mostly unidentifiable. More commonly, fossil wood constitutes the only plant material present, as a consequence of its robust and sturdy character.

The nature of the contact between the Scarborough Formation and overlying Scalby Formation has stimulated much speculation over the years. Although there is no doubt of its erosive nature (at least locally), the time-interval that this junction represents is rather more conjectural. Two main theories exist to explain the nature of this contact, one asserting that sedimentation from the Scarborough to Scalby Formation was continuous (Fisher and Hancock 1985; Riding and Wright 1989), the other stating that there was a significant break in deposition which lasted an interval equating to the greater part of the Bathonian (Leeder and Nami 1979; Hogg 1993; Hesselbo and Jenkyns 1995). Currently the latter option, that a major stratigraphical gap occurs at the base of the Scalby Formation, is thought to better answer the debate and is thus referred to here (Text-fig. 2).

The Moor Grit Member forms the lower part of the Scalby Formation and comprises a maximum thickness of 8 m of coarse-grained, low-sinuosity channel sandstone. Nami (1976) showed that the overlying Long Nab Member comprises a meander-belt featuring cross-cutting point bars trending southwards, which is well exposed on the foreshore of Burniston Wyke for a distance of more than 3 km. To the south of Scarborough, the Moor Grit Member is absent, and the Long Nab Member rests directly on the Scarborough Formation (Black 1929). The amount of leaf debris increases upwards into the levee deposits on top of the point-bar system, and the succeeding shales and sandstones of the Long Nab Member contain marine palynomorphs (Hancock and Fisher 1981). The succession is interpreted as representing a swamp, across which there were occasional incursions of sea water (Kantorowicz 1990), terminating the record of deltaic deposition in Yorkshire. The Scalby Ness Plant Bed, discussed below, occupies a stratigraphical position above the meander-belt sandstone unit, in a sequence of level-bedded overbank deposits within which discrete river channels are found.

3. PLANT BED LOCALITIES

This chapter summarizes the environment of deposition and outlines the stratigraphical context of the four chosen plant beds. A comprehensive list of the plant species found at each locality is given in Table 1 at the end of this section. Maps and sedimentary logs are included below, a legend for which is provided in Text-figure 3.

Hasty Bank Plant Bed (NZ 567 036)
Before visiting Hasty Bank, permission must be sought from the Head Forester at the Forestry Commission, 9, Roseworth, Great Broughton, Middlesbrough TS9 7EN. The steep, grassy, north-facing bank exposes the lowermost Saltwick Formation, which rests upon black marine shales of the Dogger Formation. From a car park situated off Clay Bank, the locality can be reached by walking up a steep track off the main road and through a wooded area. Access points to the relevant paths are indicated in Text-figure 4.

Hasty Bank was established as an important plant fossil locality in 1929 by Black and has since undergone repeated examination, perhaps most notably by Hill (1974*a*, 1974*b*), Hill and van Konijnenburg-van Cittert (1973) and Spicer and Hill (1979). Since its discovery, some 75 species have been collected from the site, primarily from a section of argillaceous overbank sediments (Text-fig. 5) which extends laterally over a distance of some 100 m (Hill 1974*b*). The best preserved plant remains occur adjacent to a large channel sandstone body which dominates the exposure. The small areas of previous palaeobotanical excavation which dot the bank may prove useful in keying into the schematic sedimentary section shown in Text-figure 5. Recently, some of the best fossils have been obtained from a siltstone bed which overlies an erosion surface (marked 'B' in Text-fig. 5), linked to the downcutting of the channel body.

Important palaeoecological studies have been carried out on the locality, most notably by Harris (1964, pp. 129–133) who postulated that the pteridosperm *Pachypteris papillosa* (Thomas and Bose) Harris grew under a marine influence in a tidal location reminiscent of today's mangroves. This inference is based primarily upon the succulent leaves and axes of the species, and its repetitive association with marine microfossils (Muir 1964).

Hayburn Wyke Plant Bed (TA 010 970)
The Hayburn Wyke Plant Bed occupies a stratigraphical position referable to the upper Saltwick Formation (Text-fig. 6). The unit dips very gently southwards in the wyke, bringing the overlying marine Eller Beck

LITHOLOGY

Pebble bed

Sandstone

Siltstone

Clay

Limestone

Oolitic limestone

Nodular siderite

Sideritic ironstone

SEDIMENTARY STRUCTURES

Ripple marks

Cross bedding

Cross lamination

Wavy and flaser bedding

Parallel lamination

Soft sediment deformation

Rip up clasts

FLORA

Rootlets

Fossil wood

Plant remains

GRAIN-SIZE PROFILE

clay
silt
very fine
fine
medium
coarse

FAUNA

Bivalves

Crinoids

Echnoid spines

Ooids

Serpulid
worm casts

TRACE FOSSILS

General bioturbation

Diplocraterion

Planolites

Thalassinoides

Skolithos

LOCALITY MAPS

P   Car park

Path

Point of access
to locality

Marks position
of plant bed

Houses/built-up area

Outcrop

Heath

Woodland

TEXT-FIG. 3. Key to sedimentary logs and locality maps.

Formation close to beach level at Iron Scar. The beach itself is rocky and rapidly overrun by a rising tide, and it is therefore advised that this excursion is carried out on a falling tide. The wyke can be approached from the north or south along the cliff-top path, or past the Hayburn Hotel and through a wooded area (Text-fig. 7).

Exposures of the Saltwick Formation in Hayburn Wyke are rather scattered and are commonly overgrown with vegetation, or covered by land-slip. Nevertheless, several exposures exist from which useful sedimentological information and varied plant fossils can be obtained. Immediately to the left of the path which leads to the beach (highlighted by an arrow on Text-fig. 7) is a number of thin, plant-bearing layers within overbank sediments. These beds occur below a prominent channel sandstone which forms a small waterfall near the path to the wyke. To the

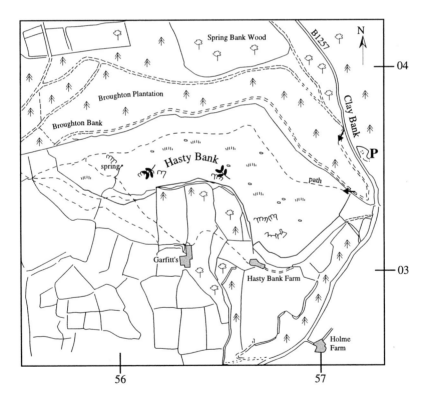

TEXT-FIG. 4. Locality map of Hasty Bank (NZ 567 036). Car parking (P) and access points are indicated, as well as the main plant-bearing exposure. For key to symbols see Text-figure 3. Grid lines are 1 km apart.

south, this channel body can be traced with care to a level of crevasse-splay deposits, beneath which further plant assemblages exist within argillaceous sediments of flood-plain origin, and thin sheet-sandstones. Here, in the centre of the wyke, these plant-yielding strata are encompassed by the Hayburn Wyke Plant Bed. The sedimentary logs shown in Text-figure 6 record these beds. The range of species and preservation states varies considerably in the plant bed interval, largely a reflection of the nature of the sediments within which the plant material is contained. In general terms, the sandy units include more fragmentary plant material which is less taxonomically diverse; much of the material is charcoalified. In contrast, the finer-grained beds contain better preserved, more abundant, and diverse plant remains.

Gristhorpe Plant Bed, Cayton Bay (TA 081 844)
The Gristhorpe Plant Bed contains one of the most diverse and abundant plant assemblages found in the Yorkshire Jurassic. The bed was brought to the attention of palaeobotanists by the attempts of Thomas (1925) to reconstruct *Caytonia*. His work was based on the repetitive association of the leaf *Sagenopteris phillipsii* (Brongniart) Presl and the pteridosperm fructifications *Caytonia sewardii* Thomas (female) and *Caytonanthus arberii* (Thomas) Harris (male), in his belief that *Caytonia* might

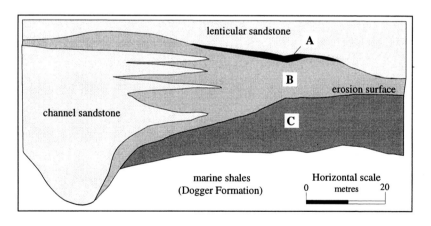

TEXT-FIG. 5. Schematic cross section of Hasty Bank (NZ 567 036), looking south (Hill and van Konijnenburg-van Cittert 1973). A large channel sandstone and a thick sequence of flood-plain sediments comprise this portion of the lowermost Saltwick Formation, overlying marine shales of the Dogger Formation. Plant fossils are abundant and well preserved in the flood-plain deposits, particularly those beds marked A, for dark-grey clay; B, for siltstone; and C, for light-grey clay. This diagram shows a vertical exaggeration of × 4 the horizontal scale.

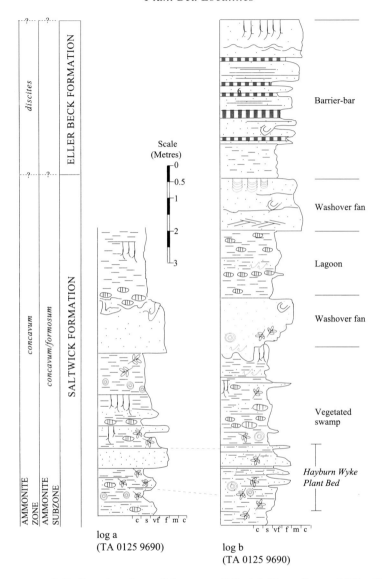

TEXT-FIG. 6. Sedimentary logs of the upper Saltwick Formation and Eller Beck Formation, recorded at Hayburn Wyke (TA 010 970) (Morgans 1997). The stratigraphical position of the Hayburn Wyke Plant Bed is indicated on log b. Text-figure 7 shows the location of these logs within Hayburn Wyke. The succession within the wyke dips gently southwards, and the Eller Beck Formation reaches beach level at Iron Scar. See Text-figure 3 for a key to the symbols.

be related to the angiosperms. Although Harris (1951) subsequently questioned such a relationship, the debate and discussion generated by Thomas's work succeeded in establishing the importance of this locality.

The Gristhorpe Plant Bed is located at Yon's Nab, the headland to the south of Cayton Bay (Text-fig. 8). From the car park in the centre of Cayton Bay, Yon's Nab can be reached by descending to the cliff foot and walking across the sandy beach and a stretch of boulders. Next, one should proceed along the path leading over the undercliff immediately west of Yon's Nab. Alternatively, Yon's Nab can be approached along the cliff-top path and down a steep and uneven route over slumped ground to beach level. This slumped area reflects the position of a major fault known as the Red Cliff Fault, which was active during the Mid Jurassic. Examination of the Gristhorpe Plant Bed is best done on a falling tide, particularly as much of the best exposure is on the foreshore and can only be seen within a few hours of low tide. The plant bed can be traced to the headland where it is above the high water mark, but at this point the plant fossils tend to be

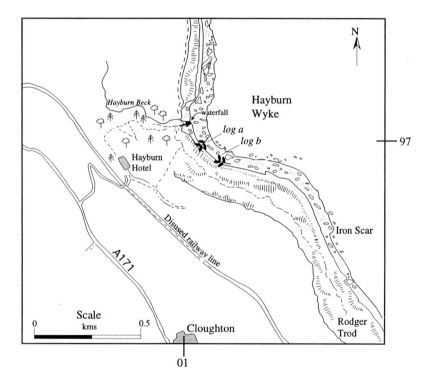

TEXT-FIG. 7. Locality map of Hayburn Wyke. The positions of the two sedimentary logs shown in Text-figure 6 are indicated. See Text-figure 3 for key to symbols.

more comminuted. It is advised that hard hats are worn for study of this part of the cliff section.

Low tide reveals a sequence ranging from the top of the Sycarham Member to the lower Scalby Formation at Yon's Nab (Text-fig. 9 records most of this sequence). The Gristhorpe Plant Bed occurs near the base of the Gristhorpe Member, which is exposed as a laterally continuous exposure more than 100 m long. The unit rests directly upon the Lebberston Member, which at this locality is clearly divisible into its two constituent parts: the Millepore Bed and the Yons Nab Beds. A 5 m thick coarsening-upward succession comprises the Yons Nab Beds, containing a fully marine fauna and displaying intense bioturbation which has destroyed the original sedimentary structures. The units as a whole represent a period of shallowing, substantiated by the upward increase in plant material. On the southern side of the headland a large channel sandstone cuts down into and truncates these beds. To the north of the headland, this channel is equivalent to a crevasse-splay sandstone that occurs as a prominent ledge running around the base of the cliff and dipping down on to the foreshore. Crevasse-splays form during times of flood, when river margins are breached and crevasses are created in the levees, causing a

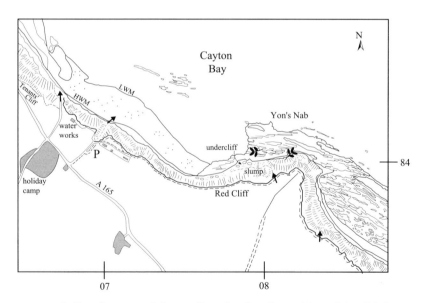

TEXT-FIG. 8. Locality map of Cayton Bay, showing the position of the Gristhorpe Plant Bed. Car Parking (P) and access points are indicated. Grid lines are 1 km apart. HWM and LWM correspond to High Water Mark and Low Water Mark respectively.

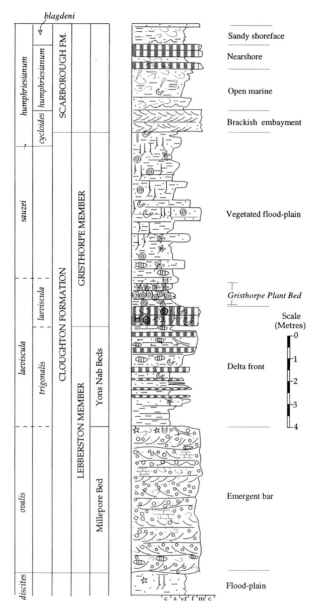

TEXT-FIG. 9. Sedimentary log displaying the sequence exposed at Yon's Nab, Cayton Bay (TA 081 844) (Morgans 1997). The stratigraphical position of the Gristhorpe Plant Bed is indicated.

lake to form on the floodplain. It is above this level of channel development and crevasse splaying that the Gristhorpe Plant Bed occurs.

The plant bed is situated between two crevasse-splay sandstones and can be roughly divided into three parts. At the base lies a soft, grey to white, clay-rich layer which yields some of the best plant fossils. Above, an indurated clay-rich siltstone occurs from which further fossil plants can be collected. Thin, fairly laterally extensive layers of compressed coalified and pyritized wood exist within this siltstone bed. Overlying the siltstone bed an iron-rich sequence occurs which is nodular in part. The plant bed attains a thickness of approximately 1 m. As the thickness and character of the sediment varies laterally, the above summary should be considered as a guide only. The composition and degree of fragmentation of the plant remains also varies significantly in a lateral sense. Pockets of exceptionally preserved leaves, inflorescences and fructifications are commonly found in

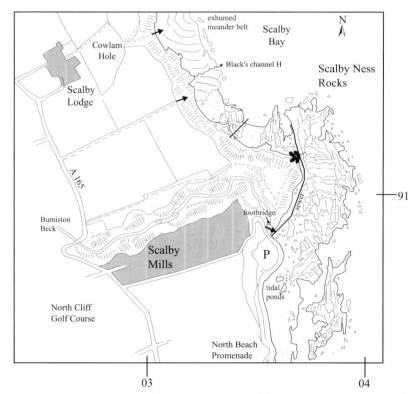

TEXT-FIG. 10. Locality map of Scalby Ness. Car parking and access points to reach the Scalby Ness Plant Bed are indicated. Grid lines are 1 km apart.

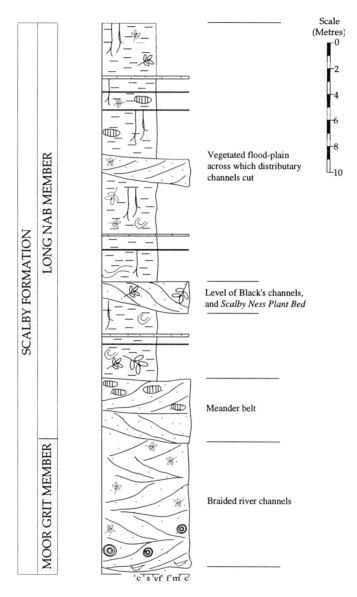

TEXT-FIG. 11. Simplified sedimentary log of the Scalby Formation, recorded at Burniston Wyke (TA 036 911) (modified from Alexander 1992). The Scalby Ness Plant Bed occurs within overbank deposits which are laterally equivalent to the level of Black's channels, shown on the log.

pale grey to white clays, closely adjacent to isolated areas of highly fragmentary material which is commonly enclosed within medium to dark grey siltstones. In many places loose boulders conceal the plant bed on the foreshore, requiring their removal for a full examination. However, enough of the plant bed is usually exposed to gain a general impression of the flora.

The plant bed is capped by an approximately 0.5 m thick micaceous sandstone displaying 'U'-shaped *Diplocraterion* burrows, indicating a marine influence during its deposition. The remaining portion of the Gristhorpe Member is composed of argillaceous flood–plain sediments intercalated with thin crevasse-splay sandstones, comprising a 9 m thick sequence in total. Succeeding these beds is the Scarborough Formation, a 5 m thick unit of marine fine-grained sandstones, carbonaceous siltstones, shelly limestones and iron-rich silty sandstones (Rawson and Wright 1992). The overlying Scalby Formation marks a return to non-marine deposition within channels and on flood–plains on the delta-plain. Fragmentary, commonly charcoalified plant debris occurs within these sediments, and includes well preserved fragments of fossilized wood.

Scalby Ness Plant Bed (TA 037 911)
This locality can be reached most easily by parking at Scalby Mills (TA 036 907) and proceeding across a footbridge to the southern end of Scalby Ness Rocks (Text-fig. 10). From here a stretch of rocky platform and beach boulders must be traversed in a northerly direction before the best exposure of the plant bed is reached. Alternatively, one may approach the locality from the north, descending to the beach in Scalby Bay. There are several points of access from the cliff-top path in this area, as shown in Text-figure 10. One such path leads from Scalby Lodge to Cowlam Hole. A falling tide is necessary to see the plant bed at its best.

A substantial accumulation of fragmentary plant material occurs at Scalby Ness, representing around 50 different plant species. Many of these taxa are rare, however, and one can usually expect to collect in the region of ten species. The bed is famed for the abundance of well preserved *Ginkgo huttonii* (Sternberg) Heer leaves. The plant remains are found within laminated shales and silts, immediately above the meander belt sandstone which occupies the foreshore of Scalby Bay (Nami 1976; Alexander 1992). Within these level-bedded, plant-bearing strata, many localized river channels developed which were first described by Black (1929), and are referred to according to Black's assignment A through H. The nearest to Scalby Ness is Black's channel H, south of Cowlam Hole (Text-fig. 10). The Scalby Ness Plant Bed thus occurs within silty and sandy overbank deposits which are brought to beach level at Scalby Ness by a gentle seaward dip. The beds from which the plant fossils are found fall within the lower Long Nab Member of the Bathonian Scalby Formation (Text-fig. 11).

TABLE 1. List of species from the Hasty Bank, Hayburn Wyke, Gristhorpe and Scalby Ness plant bed localities. Key: A=abundant; C=common; R=rare. The prefix L reflects the local occurrence of a species.

	Hasty Bank	Hayburn Wyke	Gristhorpe	Scalby Ness
Bryophyta				
Hepaticae				
Hepaticites arcuatus	–	–	R	–
Sphenophyta				
Equisetales				
Equisetum columnare	A	R	–	–
Pteridophyta				
Marattitales				
Marattia anglica	C	–	–	–
Angiopteris blackii	LC	–	–	–
Osmundales				
Todites princeps	–	R	LC	–
Todites thomasii	–	–	R	–
Todites williamsonii	–	–	A	–
Todites denticulatus	–	–	R	–
Cladophlebis harrisii	R	–	–	–
Cladophlebis denticulata	–	–	A	–
Osmundopsis hillii	R	–	–	–
Osmundopsis sturii	–	–	R	–
Filicales				
Phlebopteris polypodioides	–	–	LC	–
Phlebopteris woodwardii	–	R	R	–
Matonia braunii	–	R	R	–
Clathropteris obovata	LC	R	–	–
Dictyophyllum rugosum	R	–	C	–
Klukia exilis	–	R	LC	–
Coniopteris bella	–	–	R	C
Coniopteris simplex	–	A	–	–
Coniopteris hymenophylloides	R	C	A	R
Coniopteris murrayana	R	R	R	–
Eboracia lobifolia	–	–	LC	–
Dicksonia mariopteris	–	–	R	–
Dicksonia kendallii	LC	–	–	–
Aspidistes thomasii	–	–	R	–

TABLE 1 (*contd*)

	Hasty Bank	Hayburn Wyke	Gristhorpe	Scalby Ness
Gymnospermae				
Caytoniales				
Sagenopteris phillipsii	–	–	C	–
Sagenopteris colpodes				
large form	C	–	–	–
Sagenopteris colpodes				
small form	–	–	LC	–
Caytonia kendallii	R	–	–	–
Caytonia sewardii	–	–	LC	–
Caytonia nathorstii	–	–	LC	–
Caytonanthus arberi	–	–	R	–
Caytonanthus oncodes	–	–	R	–
Pteridosperms				
Pachypteris papillosa	A	–	–	–
with *Pteroma thomasii*	LC	–	–	–
Pachypteris lanceolata	R	R	–	–
Possible Pteridosperms				
Stenopteris williamsonis	–	–	R	–
Possible Pteridosperms				
or Cycadales				
Ctenozamites cycadea	LC	–	–	–
Ctenozamites leckenbyi	–	–	R	–
Cycadales				
Nilsonia compta	–	R	A	–
with *Beania gracilis,*	–	–	C	–
Androstrobus manis	–	–	R	–
and *Deltolepis crepidota*	–	–	R	–
Nilsonia syllis	R	R	–	–
Nilsonia kendallii	A	–	–	–
Nilsonia tenuinervis	C	–	R	R
Nilsonia tenuicaulis	–	–	LC	–
Paracycas cteis	R	–	–	–
Pseudoctenis oleosa	R	–	–	–
Pseudoctenis lanei	C	–	R	–
with *Androstrobus prisma*	R	–	–	–
Ctenis kaneharai	LC	–	–	–
Ctenis sulcicaulis	–	–	LC	–
Bennettitales				
Nilssoniopteris vittata	LC	–	A	–
with *Williamsoniella coronata*	–	–	C	–
Nilssoniopteris major	–	–	R	–
Anomozamites nilsonii	–	–	LC	–
with *Cycadolepis stenopus*	–	–	R	–
Pterophyllum thomasii	–	R	R	–

	Hasty Bank	Hayburn Wyke	Gristhorpe	Scalby Ness
Gymnospermae				
Bennettitales				
Zamites gigas	–	C	–	–
with *Williamsonia gigas*	–	R	–	–
and *Weltrichia sol*	–	R	–	–
Otozamites beanii	–	–	R	–
Otozamites graphicus	R	R	R	C
Otozamites gramineus	R	R	–	–
Otozamites tenuatus	–	R	–	–
Ptilophyllum pecten	–	–	LA	–
with *Williamsonia leckenbyi,*	–	–	R	–
Weltrichia pecten	–	–	R	–
and *Cycadolepis nitens*	–	–	R	–
Ptilophyllum pectinoides	A	–	–	–
with *Williamsonia hildae,*	R	–	–	–
Weltrichia whitbiensis	R	–	–	–
and *Cycadolepis hypene*	C	–	–	–
Ptilophyllum hirsutum	–	–	–	R
Ginkgoales				
Ginkgo sibirica	R	–	–	–
Ginkgo whitbiensis	R	–	–	–
Ginkgo digitata	–	–	LC	–
Ginkgo longifolius	–	–	R	–
Ginkgo huttoni	–	R	–	C
Baiera furcata	–	R	–	–
Sphenobaiera gyron	C	–	–	–
Sphenobaiera pecten	–	R	–	–
Eretmophyllum whitbiensis	R	–	–	–
Eretmophyllum pubescens	–	–	LC	–
Czekanowskiales				
Czekanowskia blackii	–	–	–	LC
Czekanowskia furcula	–	R	–	–
Czekanowskia microphylla	–	–	R	–
Czekanowskia thomasii	–	–	LC	–
Solenites vimineus	R	–	C	–
with *Leptostrobus cancer*	–	–	R	–

TABLE 1 (*contd*)

	Hasty Bank	Hayburn Wyke	Gristhorpe	Scalby Ness
Gymnospermae				
Coniferales				
Brachyphyllum mamillare	C	C	LC	C
with male cones attached and	R	R	R	R
Araucarites phillipsii (female)	R	–	R	–
Brachyphyllum crucis	C	–	–	–
with male cones attached and	R	–	–	–
Hirmeriella sp.	R	–	–	–
Elatides thomasii	C	–	–	–
with male and female cones	R	–	–	–
Elatides williamsonii	–	–	A	–
with male and female cones	–	–	C	–
Elatocladus laxus	–	–	C	–
Elatocladus setosus	–	R	–	–
Elatocladus zamioides	–	–	R	–
Marskea jurassica	R	–	–	–
Pagiophyllum insigne	–	–	R	–
Lindleycladus lanceolatus	–	R	–	–
Cyparissidium blackii	–	–	–	C
with *Scarburgia hillii*	–	–	–	R
and *Pityanthus scalbiensis*	–	–	–	R
Pityocladus scarburgensis	–	–	–	LC
Bilsdalea dura	R	–	LC	–
Geinitzia rigida	–	R	–	–

4. INTRODUCTION TO SYSTEMATICS

Fossil plants are frequently found as isolated fragments of the original plant to which they belonged. As a consequence, the system of taxonomic classification adopted by palaeobotanists for fossil plants differs from that used for living plants. Unlike Linnaean binominal nomenclature for living plants which uses one name for a complete plant, the classification scheme for fossil plants uses different taxa to refer to separate plant organs or combinations of organs. Thus several different taxon names may represent what are in fact individual parts of a single plant. In the Yorkshire Jurassic flora this taxonomic disparity is well exemplified by a set of five taxa which probably comprise parts of the same organism: *Ptilophyllum pectinoides* (Phillips) Halle for the leaves, *Williamsonia hildae* Harris for the female flower, *Weltrichia whitbiensis* (Nathorst) Harris for the male flower, *Cycadolepis hypene* Harris for the scale leaves, and *Bucklandia pustolosa* for the stem. Moreover, taxa for pollen grains and detached roots are possible as well. In contrast, there are sparse examples where a taxon refers not only to an individual plant organ but also to a plant as a whole (as in living plants). One such example is *Elatides williamsonii* (Lindley and Hutton) Nathorst which applies not only to leafy shoots and male and female cones found attached to them, but also to the entire Jurassic conifer. In the paragraphs below, the instances where several taxa belong to a single plant are clearly indicated.

Below, we discuss briefly the main characters of each group of fossil plants occuring in the Yorkshire Jurassic and list the main species, providing, where necessary, a key to distinguish between the various (mainly leaf) taxa. In each case, a photograph and/or a line drawing of the species is given to aid in identification. The specimen numbers mentioned in the Plate explanations and Text-figure captions refer to the collection of the Laboratory of Palaeobotany and Palynology at Utrecht when they have the prefix 'S' and to that of The Natural History Museum, London when they have the prefix 'N'. The guide is limited to the more common taxa and omits those species seldom found. Unless otherwise stated, a comprehensive review of the fossils described can be found in Harris (1961, 1964, 1969, 1979) and Harris *et al.* (1974). The classification scheme of Taylor and Taylor (1993) has been used. It is of note that several of the fossil plants discussed below are attributed to extant groups, such as Equisetales which includes the living horsetails, and even to extant genera, for example *Marattia, Angiopteris* and *Ginkgo*. Conversely, many of the groups found in the Yorkshire Jurassic are now extinct, such as the pteridosperms (seedferns) and the Bennettitales.

Key to general groups of plants
In order to come to one of the general categories of plant recognized in the succeeding pages (Bryophyta, Sphenophyta, Pteridophyta or gymno-sperms), the following general key may be of help.

1. Vegetative remains thalloid, with a clear midrib.....Bryophyta-Hepaticae
   Vegetative remains non-thalloid..2
2. Vegetative remains divided into nodes and internodes........Sphenophyta
   Vegetative remains not divided into nodes and internodes....................3
3. Vegetative remains at least once pinnate; leaves always in one plane; fertile organs, known as sporangia, attached to the surface of a (modified) leaf ....Pteridophyta
   Vegetative remains entire or pinnately divided; leaves can be in one plane but also occur spirally arranged or in bundles on short shoots.....4
4. Leaves pinnate..5
   Leaves not pinnate...7
5. Leaves thick with a leathery texture..................................Pteridosperms
   Leaves thinner,without a leathery texture 6
6. Veins simple, parallel, or sometimes (*Pseudoctenis*) forked, or even anastomosing (*Ctenis*); in the latter two cases leaves with a strongly decurrent base ....Cycadales
   Veins forking; leaves never with a strongly decurrent
   base ....Bennettitales
7. Leaflets showing net venation ....Caytoniales
   Leaves or leaflets without net venation...8
8. Leaves undivided, often scale- or needle-like; veins never ending in the margin but converging towards the apex ....Coniferales
   Leaves on short shoots, usually divided into segments; veins partly ending in the margin, partly near the apex..9
9. Leaves with broad segments; shed singly ....Ginkgoales
   Leaves with long, linear segments, after shedding remaining attached in groups to the short shoots..Czekanowskiales

Sometimes, however, the fossil plant fragments are so small that they do not provide enough information to enable determination using the general key. The Hepaticae (liverworts), with their thin thallus and distinct midrib, and the Sphenophyta (horsetails) which have stems that are divided into nodes and internodes and that are usually ribbed, are normally easy to distinguish. The problem with the Pteridophyta (=ferns) is that the sporangia are the main distinguishing character but as often as not the fossil fern fragments are sterile. In this case the best criterion to use is the pinnate foliage (often resembling that of living ferns) that is always in one plane. The venation is either pecopterid, sphenopterid or neuropterid (for explanation of these terms, see the glossary and Text-fig. 43) and never parallel. Although pinnate leaves in one plane occur in the Gymnospermae (Pteridospermae, Cycadales and Bennettitales), their venation usually distinguishes them from the Pteridophyta. Only the pteridosperms (=seedferns) may pose problems, as they may have a venation like that of the ferns. However, their leaves are always rather thick and not thin as in the ferns.

It is also difficult to distinguish between some of the gymnosperm groups on leaf morphology alone, especially between the Cycadales and the Bennettitales where the cuticle is the main character for determination of the leaves (e.g. *Nilsonia tenuinervis* Seward and *Nilssoniopteris vittata* (Brongniart) Harris are on first view alike). They can be distinguished by a small difference in venation, and by their entirely different cuticles and fructifications. An attempt to distinguish between the leaf remains of these groups is given in the above key.

5. BRYOPHYTA

No moss-like fossils have been found so far in the Yorkshire Jurassic plant beds. Liverworts (Hepaticae) are found, but they are rare fossils and in many cases the thallus is reduced to a very thin substance. Of the few species which have been collected to date, only *Hepaticites arcuatus* (Lindley and Hutton) Harris occurs with any regularity. Fragments of this species can be found in the Gristhorpe Plant Bed, revealing a dichotomously branched thallus composed of a delicate lamina often showing its cellular details when immersed in oil, and a pronounced midrib up to 1 mm wide (Text-figs 12, 13A). Occasionally, a specimen is so large that it looks as if it has been preserved *in situ* creeping over a soil surface as liverworts do today.

TEXT-FIG. 12. Schematic drawing of *Hepaticites arcuatus* (Lindley and Hutton) Harris; × 2.

TEXT-FIG. 13. A, *Hepaticites arcuatus* (Lindley and Hutton) Harris; S. 1001; dichotomously branched thallus; Gristhorpe Plant Bed; × 1. B–C, *Equisetum columnare* Brongniart. B, S. 2369; typical compression; Hasty Bank; × 1·5. C, S. 1347; cast of an erect stem; coast near Hayburn Wyke; × 1

6. SPHENOPHYTA

Today, horsetails (Equisetales) are fairly inconspicuous plants, but in the past they were large and important plants, e.g. the giant Carboniferous horsetail *Calamites* which reached heights of 10 m! The Mesozoic horsetails were comparatively small, but still quite substantial by today's standards, as stems up to 150 mm wide have been found. At certain times in the Mesozoic, horsetails formed an important constituent of the flora, especially in marsh environments. This does not, however, seem to be case in the Yorkshire Jurassic plant beds. *Equisetum columnare* Brongniart is the only common species (Text-figs 13B, 14), but it is restricted to plant beds at the base of the Saltwick Formation at Hasty Bank. Here it is particularly abundant, and small fragments are ubiquitous. Interestingly, it is a rare species in the contemporaneous strata at Hayburn Wyke (Text-fig. 13C) where occasionally, erect stem fragments can be found.

Equisetalean axes are characterized by nodes carrying whorls of partly adnate leaves (so-called leaf sheaths) with intervening smooth or longitudinally marked internodes. *Equisetum columnare* Brongniart is no exception. The unbranched stems are typically 40–50 mm wide near the base (the widest specimen known measured 65 mm) and 20 mm wide near the apex. Judging by these finds it seems probable that the species was quite tall (erect stem fragments up to 300 mm long have been found). The internodes are smooth, but because the nodes are rather crowded, the internodes themselves are often almost completely covered by 20 mm long

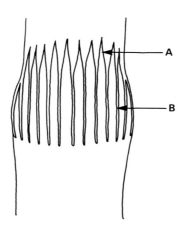

TEXT-FIG. 14. Schematic drawing of a node of *Equisetum columnare* Brongniart; A free parts of the leaves; B, leaf sheath with commissural furrows; × 1.

leaf sheaths. Each sheath consists of adnate raised leaf segments, with commisural furrows between. These furrows form 2–5 mm long acute teeth, which diverge slightly away from the stem making them prone to disarticulation (see Text-fig. 14).

7. PTERIDOPHYTA

Living ferns are divided into the Eusporangiatae (with the orders Marattiales and Ophioglossales), the Protoleptosporangiatae (Osmundales) and the Leptosporangiatae (Filicales or true ferns, and the waterfern orders Salviniales and Marsileales). The fossil record of the Ophioglossales dates back only as far as the Tertiary, but the other groups are considerably older: for example, marattialean remains have been encountered in the Carboniferous, and the oldest osmundalean remains date from the Permian.

Most fern fragments found in the Yorkshire Jurassic, can be attributed to living fern orders, families or even genera. Broadly speaking, fern specimens can be divided on a fertile/sterile basis: fertile ferns are common, and are placed in genera such as *Osmundopsis* and *Todites*; in contrast, sterile specimens are included in a form genus such as *Cladophlebis,* a genus for fern leaves with pecopterid pinnules. Whereas fertile remains are usually easy to identify, sterile material is a little more difficult and, therefore, a key is provided on page 41 to help discern between *Cladophlebis*-type leaves. It should be noted that the *Cladophlebis*-type of sterile leaf may be attributed to various fern families, such as Osmundaceae, Schizaeaceae and Dicksoniaceae. Other ferns have a very different sterile leaf shape and cannot not be included in this key (e.g. the marattialean ferns, members of the Matoniaceae and Dipteridaceae, and the various *Coniopteris* species). A general key to distinguish between these major types of fossil sterile fern leaves is given here.

1. Pinnae large; secondary veins anastomising, forming a network of vein
   meshes ....Dipteridaceae
   Pinnae large; secondary veins single or forked once ....Marattiaceae
   Pinnae divided into pinnules 2
2. Pinnules sphenopterid in shape and venation ....*Coniopteris*
   Pinnules pecopterid in shape and venation..................*Cladophlebis*-type
   Pinnules linear, long (at least twice as long as broad) with a strong
   midrib and almost perpendicularly arising lateral veins ......Matoniaceae

Simple keys to distinguish between the leaves within some of the families (e.g. Marattiaceae, Dipteridaceae) or genera (*Coniopteris*) are given in the relevant sections below.

In the Yorkshire Jurassic plant beds, members of the following groups have been found: Marattiales (family Marattiaceae), Osmundales (family Osmundaceae), and Filicales (families Matoniaceae, Dipteridaceae, Schizaeaceae, Dicksoniaceae and Thelypteridaceae). Each is discussed below.

Marattiales

At Hasty Bank, two members of this order (and of the only family Marattiaceae) are found: these can be attributed to the extant genera *Marattia* and *Angiopteris* respectively. Due to the lack of any real morphological differences from the fertile and sterile living equivalents, the fossils have been placed in these genera. As the key below reveals, the main difference between *Marattia* and *Angiopteris* is in the arrangement of the sporangia; in *Marattia*, the sporangia, arranged in two rows, are adnate, forming so-called synangia, whilst in *Angiopteris* the sporangia are arranged separately, in a sorus (see Text-fig. 16).

1. Pinnae more than 15 mm wide; sporangia fused into synangia ....
....*Marattia anglica*
Pinnae less than 15 mm wide; sporangia separately arranged in sori ....
....*Angiopteris blackii*

A complete leaf of *Marattia anglica* (Thomas) Harris has yet to be found, but pinna and leaf fragments are quite common at Hasty Bank (Harris 1961; van Cittert 1966). Pinnae can reach 300 mm long and 15–25 mm wide, their margins are entire, and the lateral veins arise almost perpendicularly to the midrib at a concentration of 10–12 per 10 mm. More than half of the specimens found are fertile (Text-fig. 15A), where

TEXT-FIG. 15. A, *Marattia anglica* (Thomas) Harris; S. 2703; fertile pinna fragment with synangia covering most of the leaf surface; Hasty Bank; × 1. B–C, *Angiopteris blackii* (van Cittert) van Konijnenburg-van Cittert; S.4758; Hasty Bank; × 2. B, fertile pinna fragment with sori (on the right) covering only a small part of the leaf surface. C, counterpart showing sori consisting of separate sporangia (see arrow).

the synangia are mainly 5–7 mm in size, although examples up to 10 mm have been found. The synangia are composed of pairs of sporangia (Text-fig. 16).

Apart from the different arrangement of the sporangia, the main distinction between pinnae of *M. anglica* and *Angiopteris blackii* (van Cittert) van Konijnenburg-van Cittert (Text-fig. 15B–C) is in the smaller size of the latter. The pinnae of *A. blackii* are about 10 mm wide and tend to occur in fragments, some up to 100 mm long, scattered throughout the enclosing sediment. To date, a complete pinna has yet to be found. Secondary veins are more crowded than in *M. anglica* (15–18 per 10 mm), and the sori normally consist of six to eight sporangia, occupying only a small part of the lamina (van Cittert 1966; Text-fig. 16).

Osmundales
As difficulties are often encountered when attempting to distinguish between sterile foliage from the osmundaceous and remains from other fern families, a key is provided below for all sterile foliage of *Cladophlebis*-type leaves. The name *Cladophlebis* is applied to those remains where no fertile material of the species is known, or cannot be attributed with certainty. This type of fossil of sterile foliage may be keyed out as follows.

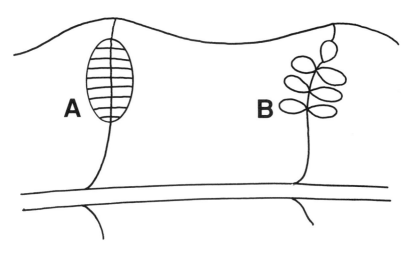

TEXT-FIG. 16. Schematic drawing of the arrangement of the sori in the marattiaceous ferns. A, synangium of *Marattia anglica* (Thomas) Harris. B, sorus of *Angiopteris blackii* (van Cittert) van Konijnenburg-van Cittert.

TEXT-FIG. 17. A, *Osmundopsis hillii* van Konijnenburg-van Cittert; V.60955; by courtesy of Dr C. R. Hill; Hasty Bank; × 2. B–C, *Cladophlebis harrisii* van Cittert; Hasty Bank. B, S. 2707; close-up of venation; × 4. C, S. 2692; general aspect of pinnules; × 1·5. D, *Osmundopsis sturii* (Raciborski) Harris; V. 63876a; Gristhorpe Plant Bed; × 1. E, *Todites denticulatus* (Brongniart) Krasser; S. 1152; Gristhorpe Plant Bed; × 1

Most fossil osmundaceous ferns have fertile foliage which resembles the sterile foliage, except that the former is covered with sporangia (e.g. *Todites princeps, T. williamsonii, T. thomasii* and *T. denticulatus*). Some examples do, however, have dimorphic sterile and fertile leaves, where the fertile leaves are greatly reduced with ultimate branches carrying groups of sporangia. They are assigned to *Osmundopsis*, a fossil equivalent of the living genus *Osmunda*. Because of this dimorphism it is often quite difficult to ascertain which sterile leaves belonged to an *Osmundopsis* species.

In Yorkshire, two *Osmundopsis* species occur: *O. hillii* van Konijnenburg-van Cittert is found at Hasty Bank and *O. sturii* (Raciborski) Harris in the Gristhorpe Plant Bed. Although the former species is exceptionally rare at Hasty Bank (Text-fig. 17A), its sterile foliage can be recognized by a distinct venation (Text-figs 17B, 18E) and is known as *Cladophlebis harrisii* van Cittert (Text-fig. 17B–C) (van Konijnenburg-van Cittert 1996). Due to the rarity of *Osmundopsis sturii* (Text-fig. 17C) in the Gristhorpe Plant Bed, the character of its sterile foliage is still unclear. There are indications that small leaves of *Cladophlebis denticulata* (Brongniart) Fontaine may be affiliated (van Konijnenburg-van Cittert 1996), but *C. denticulata* – at least the larger, clearly dentate form – is now firmly linked to *Todites denticulatus* (Brongniart) Krasser (Text-fig. 17E). So, there seems to be a possibility that *Cladophlebis denticulata* (Pl. 1, fig. 1) represents the sterile foliage of two different species!

Todites williamsonii (Brongniart) Seward (Pl. 1, fig. 2) is by far the most common of the four *Todites* species found within the Gristhorpe

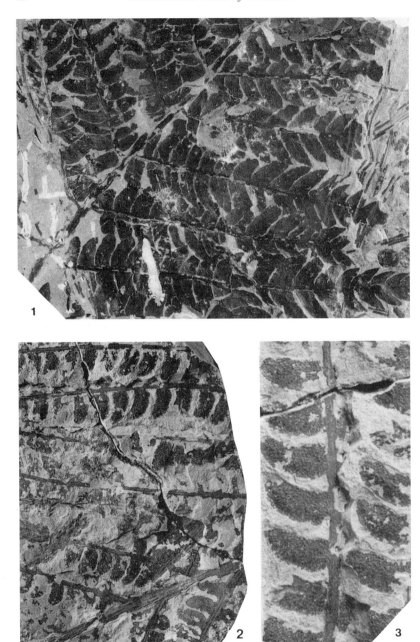

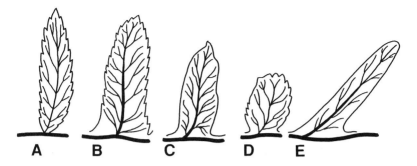

TEXT-FIG. 18. Schematic drawing of the sterile pinnule shape of the various osmundaceous ferns. A, *Todites thomasii* Harris. B, *Cladophlebis denticulata* (Brongniart) Fontaine. C, *Todites williamsonii* (Brongniart) Seward. D, *Todites princeps* (Presl) Gothan. E, *Cladophlebis harrisii* van Cittert. All × 4.

Plant Bed, Cayton Bay. It can be easily recognized by its shape, the comparatively thick main rachis, and approximate neuropterid venation with twice to three times forked lateral veins (Text-fig. 18C). Most of the specimens are fertile (Pl. 1, figs 2–3), an exception in the Yorkshire osmundaceous fossils (Harris 1961).

Todites princeps (Presl) Gothan (Pl. 2, fig. 1) is characterized by an anadromic venation (Text-fig. 18D): the only *Cladophlebis*-type leaf of this sort, as all others exhibit katadromic venation (see Text-fig. 19 and the glossary for explanation of anadromic and katadromic). The species has long, narrow leaves, 60–100 mm wide, with by far the smallest pinnules of all the osmundaceous fossils. It is locally common in the Gristhorpe Plant Bed, and good specimens have been found at Hayburn Wyke.

Todites thomasii Harris and *T. denticulatus* both show sharply toothed pinnules, but can be distinguished by the pinnule bases: those of *T. thomasii* are contracted on both sides (Text-fig. 18A), whereas the pinnules of *T. denticulatus* are attached by a broad base (Harris 1961; Text-fig. 18B). Fertile pinnules of both species differ from the sterile ones in that the

EXPLANATION OF PLATE 1

Fig. 1. *Cladophlebis denticulata* (Brongniart) Fontaine; S.2338A; Gristhorpe Plant Bed; ×1.

Figs 2–3. *Todites williamsonii* (Brongniart) Seward; S. 1172; Gristhorpe Plant Bed. 2, fertile specimen; × 1. 3, detail of fig. 2 showing fertile pinnules with sporangia; × 2.

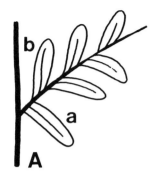 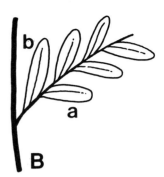

TEXT-FIG. 19. Schematic drawing of A, katadromic branching and B, anadromic branching. In both figures, 'a' is the first pinnule on the basiscopic side, and 'b' is the first pinnule on the acroscopic side.

margins of most specimens are less dentate (*T. denticulatus*, see Text-fig. 17E), whilst some are not dentate at all and instead have 'wavy' outlines (*T. thomasii*; see Pl. 2, fig. 2). The fertile *T. denticulatus* is rather common and its sterile foliage *Cladophlebis denticulata* (Brongniart) Fontaine is abundant in the Gristhorpe Plant Bed; but, as previously mentioned, there is a possibility that part of *Cladophlebis denticulata* may in fact be the sterile foliage of *Osmundopsis sturii* (Raciborski) Harris.

Filicales (Matoniaceae)
Living matoniaceous ferns are characterized by leaf morphology (Text-fig. 20A) and nature of the sori. Sori are arranged in two rows along the midrib on the under side of the leaf, and consist of a ring of ovoid sporangia around a receptacle (or placenta as it is sometimes called) which is expanded at the apex into a large peltate indusium (Text-fig. 20B).

Some fossilized matoniaceous remains have been attributed to the living genus *Matonia*, such as *Matonia braunii* (Goeppert) Harris in Yorkshire, whilst others, which differ from the living genera only in the lack of an indusium, are usually attributed to the genus *Phlebopteris*. Three members of the Matoniaceae occur in the Gristhorpe Plant Bed, and can be identified using the following key.

1. Specimens (pinnule fragments) preserved as fusain ....
....*Phlebopteris woodwardii*
    Specimens preserved as a compression (film of coal) 2
2. Vein branches all free ....*Matonia braunii*
    Vein branches anatomising ....*Phlebopteris polypodioides*

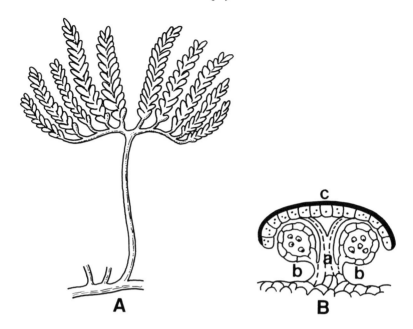

TEXT-FIG. 20. Gross morphology of leaf and sori of the Matoniaceae. A, general shape of a matoniaceous leaf. B, longitudinal section of a matoniaceous sorus showing the receptacle (a), the sporangia (b) and the covering indusium (c).

Of these three species, *Phlebopteris polypodioides* Brongniart (Pl. 2, fig. 3) is by far the most common and occurs with relative prevalence in the Gristhorpe Plant Bed. It is can be distinguished by the venation (Text-fig. 21A) which consists of a prominent midrib with lateral veins. The branches fork forwards and backwards joining with those of adjacent veins

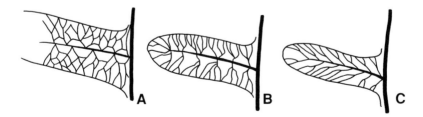

TEXT-FIG. 21. Venation patterns. A, *Phlebopteris polypodioides* Brongniart. B, *Phlebopteris woodwardii* Leckenby. C, *Matonia braunii* (Goeppert) Harris. All × 2.

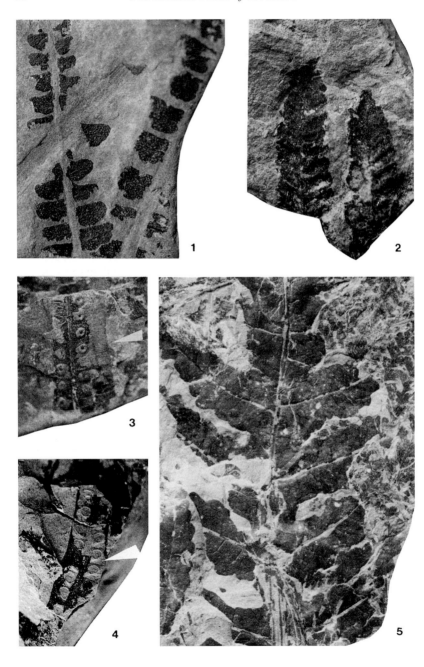

to form primary arches, which give off two to five outer veins running nearly transverse to the margin. The veins usually fork and the branches anastomose, forming elongated meshes. In fertile leaves, the sori are placed on top of the primary arches.

Phlebopteris woodwardii Leckenby and *Matonia braunii* (Goeppert) Harris (equivalent to *Matonidium goeppertii* in Harris 1961; see Harris 1980) occur only in the uppermost layer of the Gristhorpe Plant Bed and rarely at Hayburn Wyke. *Phlebopteris woodwardii* is fairly similar to *P. polypodioides,* although the outer veins arising from the primary arch usually do not anastomose (Text-fig. 21B). An interesting feature of *P. woodwardii* is that the remains only occur as small fragments of fusain, and not as coaly compressions (Harris 1961). This indicates that the remains of *P. woodwardii* are probably allochthonous and were presumably burnt before fossilization, perhaps after being struck by lightning; Harris (1981) imagined a wet season which began with high winds and electric storms. Obviously, if lightning struck prior to heavy rainfall, then the chances of fire igniting and spreading very quickly are great. Formation of charcoal by such a method may have been common in Yorkshire during the Mid Jurassic. We tend to find charcoal concentrated in discrete horizons, suggesting perhaps that heavy rains following fires washed the charcoalified remains into river systems.

Matonia braunii (Goeppert) Harris (Pl. 2, fig. 4) occurs as reasonably large, often fertile fragments in which the indusium can sometimes be seen. The lateral veins never form primary arches but fork once or twice (Text-fig. 21C).

Filicales (Dipteridaceae)

The Dipteridaceae, of which there is only one living genus, *Dipteris* in the Indo-Malaysian region, have a general frond morphology that is very similar to that of the Matoniaceae, but the lamina segments are much larger and the lateral veins branch at almost right angles to form a reticulate mesh. Fertile leaves are distinguished by the absence of an

---

EXPLANATION OF PLATE 2

Fig. 1. *Todites princeps* (Presl) Gothan; S. 7541; × 3.

Fig. 2. *Todites thomasii* Harris; S. 2312; × 4.

Fig. 3. *Phlebopteris polypodioides* Brongniart; S. 8542; note that the sori occupy only a small part of the lamina (see arrow); × 4.

Fig. 4. *Matonia braunii* (Goeppert) Harris; S.3009; note that the sori occupy most of the lamina (see arrow); × 4.

Fig. 5. *Dictyophyllum rugosum* Lindley and Hutton; S. 1163; × 0·5.

All specimens fertile, except fig. 5, and from the Gristhorpe Plant Bed.

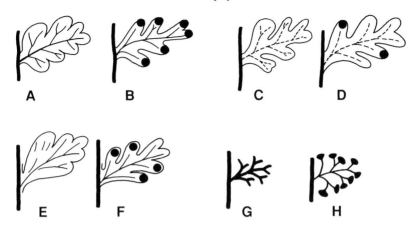

TEXT-FIG. 23. Schematic drawing of the sterile (A, C, E, G) and fertile (B, D, F, H) *Coniopteris* pinnules. A–B, *C. hymenophylloides* (Brongniart) Seward. C–D, *C. murrayana* (Brongniart) Brongniart. E–F, *C. bella* Harris. G–H, *C. simplex* (Lindley and Hutton) Harris. All × 4.

indusium. In the Mesozoic (mainly during the Triassic and Jurassic) Dipteridaceae were more common and widespread than today. The Yorkshire Jurassic reveals two species which are relatively common. *Clathropteris obovata* Ôishi is locally abundant at Hasty Bank but rare at Hayburn Wyke, and *Dictyophyllum rugosum* Lindley and Hutton is common in the Gristhorpe Plant Bed but rare at Hasty Bank. They may be distinguished as follows.

1. Pinna deeply divided into lateral segments;
   vein meshes rarely rectangular ....*Dictyophyllum rugosum*
   Pinna only slightly divided into lateral segments; vein meshes usually
   almost rectangular...*Clathropteris obovata*

Most material found associated with *Dictyophyllum rugosum* Lindley and Hutton (Pl. 2, fig. 5) is sterile, but occasionally fertile fragments can be

Fig. 1. *Coniopteris hymenophylloides* (Brongniart) Seward; S. 1198; sterile and fertile fragments; Hayburn Wyke; × 2.
Fig. 2. *Coniopteris murrayana* (Brongniart) Brongniart; S.1196; sterile fragments; Gristhorpe Plant Bed; × 2.

TEXT-FIG. 22. A, *Clathropteris obovata* Ôishi; S.7605; Hasty Bank; × 1. B, *Klukia exilis* (Phillips) Raciborski; sterile specimen from the Middle Jurassic of Iran; × 1.

found where one to four sporangia are crudely grouped in well-defined sori. Sterile specimens of *Clathropteris obovata* Ōishi (Text-fig. 22A) are more common than the fertile form, in which, unlike *D. rugosum*, distinct sori can be seen with a clear receptacle (placenta) surrounded by sporangia and solid, curved hairs.

Filicales (Schizaeaceae)

Schizaeaceae is a family of relatively primitive ferns, with large sporangia (up to 1 mm) never grouped in sori. Together with Marattiaceae, Schizaeaceae is one of the oldest fern families, possibly dating back to the Carboniferous (Taylor and Taylor 1993). Today, the family is widespread particularly in tropical and subtropical areas, where several genera and more than a hundred species occur. In the Yorkshire Jurassic, only one member of the family can be found with any regularity: *Klukia exilis* (Phillips) Raciborski (Text-fig. 22B) which is locally common in the Gristhorpe Plant Bed, and rarely found at Hayburn Wyke. The pecopterid leaves are finely divided, rachises are slender and the branching is sparse (see the key for *Cladophlebis*-type of leaves). The surfaces of the sterile pinnules are convex with curved down margins but they are never curved back under the lamina. Fertile pinnules bear large (0·5 mm) sporangia on both sides of the midrib along the simple or once forked lateral veins. The margins are often curved backwards in the fertile pinnules.

Filicales (Dicksoniaceae)

Represented by ten species (seven of which occur in the localities discussed here), Dicksoniaceae is the most diverse fern family in the Middle Jurassic flora of Yorkshire. Extant Dicksoniaceae are (sub)tropical tree ferns, and there is reason to believe that at least some of the Jurassic members of the family may have been small tree ferns.

Within the Dicksoniaceae, the four *Coniopteris* species belong to the subfamily Thyrsopterideae (with a continuous indusium), two *Dicksonia* species belong to the subfamily Dicksonioideae (with a two-valved indusium), and *Eboracia lobifolia* (Phillips) Thomas seems to be an intermediate variety with a lobed indusium (van Konijnenburg-van Cittert 1989). The two *Dicksonia* species together with *Eboracia lobifolia* have sterile foliage which approximates to the *Cladophlebis* leaf type (see key on p. 41), whilst the four *Coniopteris* species can be determined with the aid of the following key and Text-figure 23.

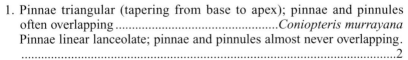

1. Pinnae triangular (tapering from base to apex); pinnae and pinnules often overlapping ....*Coniopteris murrayana*
   Pinnae linear lanceolate; pinnae and pinnules almost never overlapping.
  2
2. Segments filiform, once-veined ....*Coniopteris simplex*

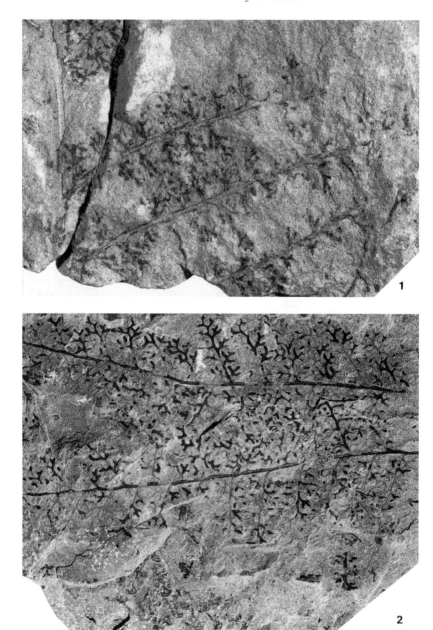

Segments broader, with two or more veins..3
3. Apices of narrower segments rounded.........................*Coniopteris bella*
   Apices of narrower segments pointed ......*Coniopteris hymenophylloides*

Sterile *Coniopteris hymenophylloides* (Brongniart) Seward (Pl. 3, fig. 1) is a very common species in the Yorkshire Jurassic flora and occurs at many localities from throughout the Ravenscar Group, including the four discussed here. The reduced fertile leaf (Pl. 3, fig. 1) is relatively rare but can be found in the Gristhorpe Plant Bed and at Hayburn Wyke.

Coniopteris murrayana (Brongniart) Brongniart (Pl. 3, fig. 2; Pl. 20, fig. 1) is a relatively rare species which is usually found in its sterile form at Hasty Bank, Hayburn Wyke and in the Gristhorpe Plant Bed. When found, fertile segments are less reduced than in *C. hymenophylloides*. A further difference between *C. hymenophylloides* and *C. murrayana* is that in the former, the first pinnule on the basiscopic side is aphlebiform (developed into filiform processes), whilst in the latter, it is the first acroscopic pinnule which is aphlebiform. This can only be seen in larger fragments.

Coniopteris bella Harris (Pl. 4, fig. 1) is characterized by rounded segments, and can be found in the Gristhorpe Plant Bed and more commonly at Scalby Ness.

Coniopteris simplex (Lindley and Hutton) Harris (Pl. 4, fig. 2) occurs at Hayburn Wyke and is distinguished by having a much reduced lamina with filiform segments.

Eboracia lobifolia (Phillips) Thomas (Pl. 5, fig. 1) is locally common in the Gristhorpe Plant Bed including fertile fragments. Its broad, often forked first basiscopic pinnule in combination with a reduced second (and sometimes third) pinnule is a good diagnostic character in the field, distinguishing it from *Dicksonia mariopteris* Wilson and Yates (Text-fig. 24A–B), which is a rare fossil found at the same locality. In the latter, the first basiscopic pinnule is enlarged, slightly stalked and divided into three lobes, whilst the second and third pinnules are seldom reduced. In general, pinnules of *Dicksonia mariopteris* are longer, narrower and with a rather more lobed margin than those of *Eboracia lobifolia*. *Dicksonia kendallii* Harris (Pl. 5, fig. 2) occurs at Hasty Bank and is locally common. Here, the first basiscopic pinnule is very broad but never lobed, and always, even in fertile leaves, sterile.

EXPLANATION OF PLATE 4

Fig. 1. *Coniopteris bella* Harris; S. 3031; fertile specimen; Scalby Ness; × 2.
Fig. 2. *Coniopteris simplex* (Lindley and Hutton) Harris; S.1187; mainly sterile specimen, except at the lower edge of the photo where there is a small fertile part; Hayburn Wyke; × 2.

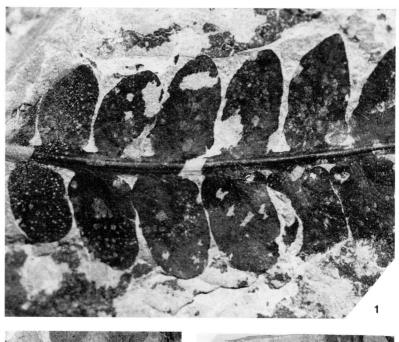

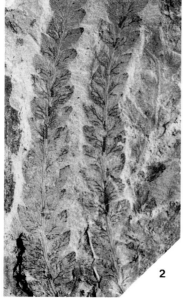

Filicales (Thelypteridaceae)
Aspidistes thomasii Harris (Text-fig. 24c) was included by Lovis (1975) in the Thelypteridaceae. Harris (1961) attributed it to the Aspideae? This slender fern with small lobed pinnules can be found in the Gristhorpe Plant Bed. Its rachises are relatively stout, the branching is crowded and most of the fragments are fertile, although the plane of cleavage of the specimens rarely exposes the sori.

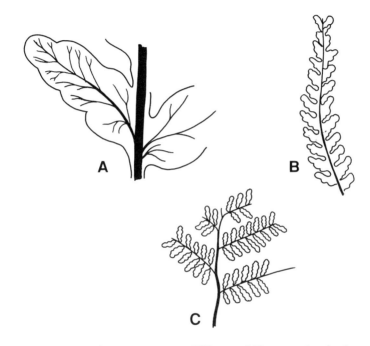

TEXT-FIG. 24. A–B, *Dicksonia mariopteris* Wilson and Yates. A, pinnule shape and venation; × 2·5. B, general pinna morphology; × 0·5. C, *Aspidistes thomasii* Harris; fragment of a frond; × 0·5 All after Harris (1961).

EXPLANATION OF PLATE 5

Fig. 1. *Eboracia lobifolia* (Phillips) Thomas; V. 63935; sterile specimen; Gristhorpe Plant Bed; × 3.

Fig. 2. *Dicksonia kendallii* Harris; S. 4752; sterile specimen, note the venation; Hasty Bank; × 2.

Fig. 3. *Sagenopteris phillipsii* (Brongniart) Presl; S. 2644; leaf consisting of four leaflets; Gristhorpe Plant Bed; × 1.

8. CAYTONIALES

As mentioned in a previous section, Thomas (1925) made the Gristhorpe Plant Bed famous through his work on the Caytoniales. He considered these plants to be early angiosperms, and indeed the leaves (belonging to the genus *Sagenopteris*) look more like angiosperm than normal gymnosperm leaves. A complete leaf consists of a petiole bearing four lanceolate leaflets (Pl. 5, fig. 3) (the leaflets are usually found detached from the petiole) showing a net venation (Pl. 6, fig. 1; Text-fig. 25). Thomas's reconstruction of the *Caytonia* plant was based on the repetitive association of *Sagenopteris phillipsii* (Brongniart) Presl with the female fructification *Caytonia sewardii* Thomas and the male *Caytonanthus arberi* (Thomas) Harris.

Caytonia plants have now been recorded from various parts of the world, but most of the taxa are from Europe. Apart from the example mentioned above, there are two other *Caytonia* plants from Yorkshire. One can be found at Hasty Bank (large leaves of *Sagenopteris colpodes* Harris with *Caytonia kendallii* Harris (Text-fig. 26E) but so far without a *Caytonanthus* species) and the other in the Gristhorpe Plant Bed (small leaves of *Sagenopteris colpodes* with *Caytonia nathorstii* (Thomas) Harris and *Caytonanthus oncodes* Harris).

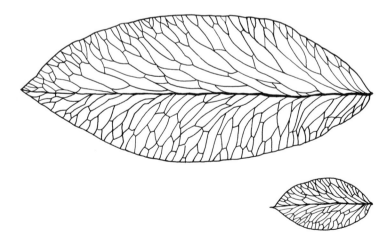

TEXT-FIG. 25. *Sagenopteris colpodes* Harris; schematic drawings of leaves of small form (left) and normal form (right); both show net venation; × 1.

Reproductive organs

Caytonia is a genus for megasporophylls with a main rachis bearing seed-containing sacs (the so-called 'fruits', which are around 4 mm in diameter) in nearly opposite pairs (Pl. 6, fig. 3). The fruits have short stalks, are globose and fleshy with a contracted 'mouth', curved back against the stalk and closed at maturity (Text-fig. 26A–B). Thomas (1925) attributed these fruits to the angiosperms, but subsequent work by Harris (1940, 1941, 1951) proved that pollination was essentially gymnospermous, and today Caytoniales is considered to be a special group within the seed ferns (Pteridospermae).

There are three different *Caytonia* species in Yorkshire: *C. kendallii* Harris (Text-fig. 26E) occurs at Hasty Bank, where it is very rare, the other two species occur in the Gristhorpe Plant Bed where they are locally common. They can be separated using the following key.

1. Mouth of fruit around 2 mm ....*Caytonia kendallii*
   Mouth of fruit around 1 mm 2
2. Main rachis smooth..*Caytonia sewardii*
   Main rachis longitudinally ribbed............................*Caytonia nathorstii*

It is almost impossible to distinguish in the field between detached fruits of the latter two species.

Caytonanthus is a genus for microsporophylls with a main rachis branching in a horizontal plane. The branches are in more or less opposite pairs and are simple or forked (Pl. 6, fig. 5). Ultimate branchlets bear a single, almost terminal, elongated synangium composed of four pollen sacs (Pl. 6, fig. 2; Text-fig. 26C–D). Two species are known from the Gristhorpe Plant Bed. They can be distinguished by the shape of their synangia (Text-fig. 26C–D): *C. arberi* (Thomas) Harris has longer and more acute synangia than *C. oncodes* Harris.

EXPLANATION OF PLATE 6

Fig. 1. *Sagenopteris phillipsii* (Brongniart) Presl; S. 2644; detail of Pl. 5, fig. 3 showing venation; × 2.

Figs 2, 5. *Caytonanthus arberi* (Thomas) Harris; × 2. 2, S. 2323, pollen sacs. 5, S. 8520; rachis with more-or-less opposite side branches and one separate pollen sac.

Fig. 3. *Caytonia sewardii* Thomas; S. 1230; several 'fruits'; × 2.

Fig. 4. *Sagenopteris colpodes* Harris; complete large leaf and fragments of small leaves; × 1.

All specimens from the Gristhorpe Plant Bed.

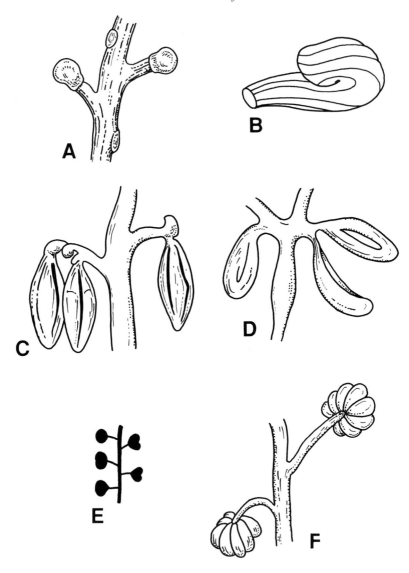

TEXT-FIG. 26. Caytonialean and pteridosperm reproductive structures. A–B, *Caytonia nathorstii* (Thomas) Harris. A, general morphology; × 1·5. B, general diagram of a fruit; × 4. C, *Caytonanthus arberi* (Thomas) Harris; × 4. D, *Caytonanthus oncodes* Harris; × 4. E, *Caytonia kendallii* Harris; schematic drawing of the general morphology; × 0·6. F, *Pteroma thomasii* Harris; part of an axis with two microsporophylls attached; × 8. A–B and F after Harris (1964).

Foliage

Caytonialean foliage is assigned to the genus *Sagenopteris*, of which three types are dealt with here, distinguishable as follows.

1. Leaflets six times (or more) as long as broad ....*Sagenopteris phillipsii*
   Leaflets two to five times as long as broad...2
2. Leaflets < 40 mm long..*Sagenopteris colpodes*
   (small form)
   Leaflets > 40 mm long (usually > 100 mm)..........*Sagenopteris colpodes*
   (large form)

Sagenopteris colpodes Harris (Pl. 6, fig. 4) is an aggregate of two very similar species which differ only in size. Large and small forms (Text-fig. 25) are imperfectly distinguishable using present evidence, but fortunately their corresponding *Caytonia* species differ considerably. In the field, however, this should not present any difficulties, as the large form occurs at Hasty Bank and the small form in the Gristhorpe Plant Bed, where it is only locally common, in contrast with *Sagenopteris phillipsii* (Brongniart) Presl (Pl. 5, fig. 3; Pl. 6, fig. 1) which is common throughout the whole plant bed.

9. PTERIDOSPERMS

Mesozoic pteridosperms are distinct from other gymnosperm remains by their pinnate, thick leathery leaves. In the localities dealt with by this field guide, five species (possibly) belonging to this very heterogenous group are present.

Key to the pteridosperm foliage

1. Leaves tripinnate, pinnae divided into narrow segments ....
   ....*Stenopteris williamsonis*
   Leaves once or twice pinnate, pinnae more robust................................2
2. Pinnae with a distinct midrib, secondary veins often invisible...............3
   Pinnae with several, equally strong, almost parallel veins.....................4
3. Leaf once pinnate, segments usually more than 5 mm wide ....
   ....*Pachypteris papillosa*
   Leaf twice pinnate, segments usually less than 5 mm wide ....
   ....*Pachypteris lanceolata*
4. Pinnules once or twice as long as broad; apex obtuse.............................
   ....*Ctenozamites cycadea*
   Pinnules twice to five times as long as broad; apex acute.......................
   ....*Ctenozamites leckenbyi*

Stenopteris
S. williamsonis (Brongniart) Harris is a pteridosperm of unknown affinities which is occasionally found in the Gristhorpe Plant Bed. The leaves are leathery, the lamina is typically tripinnate and the pinnae divided into narrow segments (Pl. 7, fig. 5), 0·5–1 mm wide, with a single median vein.

Pachypteris
The two species of this corystosperm genus present in the Yorkshire Jurassic flora (*P. papillosa* (Thomas and Bose) Harris and *P. lanceolata* Brongniart) have foliage with a leathery texture and a clear midrib. Even without the reproductive male organ *Pteroma* (*Pteroma thomasii* Harris in Yorkshire), *Pachypteris* would be treated as a pteridosperm, but the presence of *Pteroma* confirms the identification. The megasporophylls of the Yorkshire *Pachypteris* species are, however, still unknown.
   P. papillosa (Thomas and Bose) Harris (Pl. 7, figs 1–2) is known only from Yorkshire and in some localities (including Hasty Bank) it is so abundant that it forms a thin paper coal. Although typical pinnae are 12 mm by 6 mm, they can be as large as 45 mm by 10 mm! More apical pinnae are smaller of course. At Hasty Bank, *Pachypteris papillosa* is

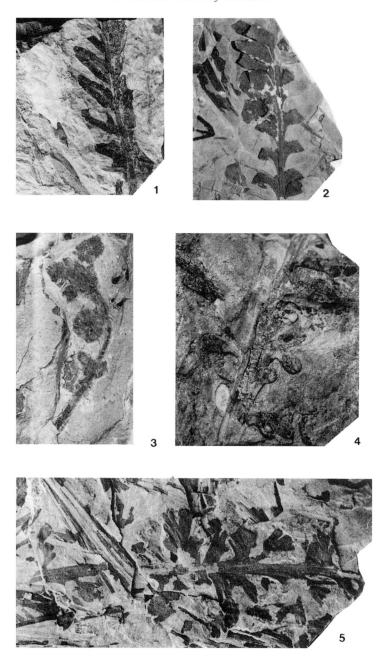

sometimes accompanied by its microsporophyll *Pteroma thomasii* Harris (Pl. 7, fig. 3). This consists of a slender, simply pinnate, approximately 30 mm long rachis with short, stalk-like branchlets terminating in an oval fertile head. These heads bear two rows of pollen sacs on their lower side (Text-fig. 26F).

P. lanceolata Brongniart (Pl. 7, fig. 4) is the type species of the genus *Pachypteris* and is a common species in Mesozoic floras in the northern hemisphere. In Yorkshire, hand specimens are not very common, although smaller fragments occur in many localities. In our four localities, *P. lanceolata* hand specimens can be occasionally found at Hasty Bank and Hayburn Wyke. Sometimes it is not easy to distinguish small specimens from *P. papillosa* but generally one can say that the pinnules are more narrow than in the latter species. Typical pinnules of *P. lanceolata* are 5–8 mm long by 2–3 mm wide, whilst large ones can be up to 15 by 4 mm.

Ctenozamites

This was formerly included in the Cycadales, but this may be incorrect as the closely related *Ptilozamites* (which has so far never been found in Yorkshire) is now regarded as a pteridosperm because of associated fructifications. The leathery texture of *Ctenozamites* leaves supports it being a pteridosperm. The lamina of the pinnae is decurrent and arises on the upper side of the rachis. The pinnae are divided into rhomboidal, triangular or falcate (curving forward) segments (i.e. pinnules), with a decurrent base.

Ctenozamites leckenbyi (Leckenby) Nathorst (Pl. 8, fig. 1) is a rare fossil in the Gristhorpe Plant Bed, with pinnules (leaf segments) which are distinctly falcate. Pinnules composing large leaves are usually 30–40 mm long by 10 mm wide, and 20 mm by 7–10 mm for smaller leaves. The apex of the pinnules is acute and often bears minute teeth. The veins are parallel, usually once forked and end in the margin near the apex. *Ctenozamites cycadea* (Berger) Schenk (Pl. 8, fig. 2) is quite a common fossil in some layers at Hasty Bank. The leaf segments are typically

---

EXPLANATION OF PLATE 7

Figs 1–2. *Pachypteris papillosa* (Thomas and Bose) Harris; × 1. 1, S. 4781; narrow pinnules. 2, S. 1058; normal pinnules.

Fig. 3. *Pteroma thomasii* Harris; S. 1527; × 2.

Fig. 4. *Pachypteris lanceolata* Brongniart; S. 3933; × 1.

Fig. 5. *Stenopteris williamsonis* (Brongniart) Harris; S. 2338; × 2.

Figs 1–4 from Hasty Bank; fig. 5 from the Gristhorpe Plant Bed.

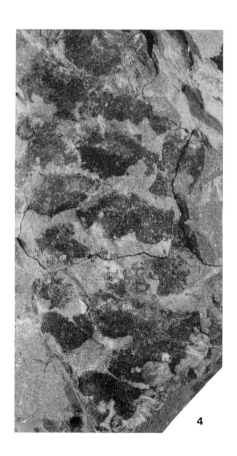

rhomboidal and 20 mm by 10 mm, but in more distal parts they become triangular and 10 mm by 7 mm. Their apex is obtuse and without teeth. The veins are parallel, once or twice forked and end in the outer margin.

EXPLANATION OF PLATE 8

Fig. 1. *Ctenozamites leckenbyi* (Leckenby) Nathorst; S. 7584; × 1.
Fig. 2. *Ctenozamites cycadea* (Berger) Schenk; S. 1425; × 2.
Fig. 3. *Nilsonia tenuinervis* Seward; S. 1293; × 1.
Fig. 4. *Androstrobus manis* Harris; S. 1370; × 2.
Figs 1, 3–4 from the Gristhorpe Plant Bed; fig. 2 from Hasty Bank.

10. CYCADALES

Living Cycadales are restricted mainly to dry, tropical areas (between 35°S and 35°N) such as south-east Africa, Australia, and South and Central America. They are a relict of a group that was widespread and diverse in the past, especially in the Mesozoic, which were typified by a thick stem covered with scale leaves, and a tuft of stiff pinnate leaves often resembling palm leaves at the apex. Male fructifications are cones consisting of closely packed, stiff microsporophylls covered on their lower side with numerous pollen sacs. Female fructifications are more loosely organized in an apical tuft, and the megasporophylls are leaf-like with two short rows of seeds on either side.

Jurassic cycads were more diverse both in leaf form (undivided or pinnate) and fructifications than extant forms, and as a consequence it is very difficult to compare them directly. In Yorkshire, we find four different leaf genera: *Nilsonia, Paracycas, Pseudoctenis* and *Ctenis*. The most common *Nilsonia* species, *N. compta* (Phillips) Bronn, can be found in the Gristhorpe Plant Bed accompanied by female fructifications (*Beania gracilis* Carruthers), male cones (*Androstrobus manis* Harris) and scale leaves (*Deltolepis crepidota* Harris). Occasionally, microsporophylls of *Pseudoctenis lanei* Thomas (*Androstrobus prisma* Thomas and Harris) can be found at Hasty Bank. Fructifications belonging to the other Yorkshire cycad species are either rare or have never been found.

Key to cycad leaves

1. Leaves undivided ....*Nilsonia tenuinervis*
   Leaves divided into segments (pinnae) 2
2. Lamina (pinnae) attached to upper edge of rachis and entirely concealing it from above 3
   Pinnae laterally inserted on the rachis 6
3. Length of leaf segments (pinnae) at least four times width...................4
   Length of leaf segments (pinnae) less than four times width................5
4. Segments curving forwards ....*Nilsonia syllis*
   Segments straight ....*Nilsonia tenuicaulis*
5. Leaf usually large, veins about 15/10 mm near rachis ...*Nilsonia compta*
   Leaf usually smaller, veins about 35/10 mm near rachis..........................
   ....*Nilsonia kendallii*
6. Pinnae arising at almost right angles, base straight ....*Paracycas cteis*
   Pinnae arising at smaller angles, lower base contracted or expanded 7
7. Venation almost parallel, without anastomoses 8
   Venation almost parallel, with anastomoses 9

8. Pinnae lanceolate, usually more than 12 mm wide, separated by a small
  distance only ....*Pseudoctenis oleosa*
  Pinnae strap-shaped, usually less than 10 mm wide, separated by a
  distance about equal to their own width ....*Pseudoctenis lanei*
9. Pinna base usually contracted ....*Ctenis kaneharai*
  Pinna base expanded...*Ctenis sulcicaulis*

Nilsonia

This is the most common and diverse genus of the Yorkshire cycads. *N.
tenuinervis* Seward (Pl. 8, fig. 3) is the only cycad with an undivided
lamina discussed here. It is difficult to distinguish it in the field from some
bennettitalean *Nilssoniopteris* species, but cuticle preparations can reveal
otherwise hidden information. Two basic differences exist: (1) the veins
never fork in *N. tenuinervis* (or indeed in any *Nilsonia* species); and (2) the
petiole in the leaf base is either absent or very short. In contrast,
Nilssoniopteris species have much longer petioles and their veins fork. *N.
tenuinervis* leaves are linear and may be more than 600 mm long, so one
usually only finds leaf fragments. The apex is acute and the lamina tapers
gradually towards the rachis base which is itself expanded. Occasionally,
the lamina is irregularly divided up to the midrib, especially in the lower
part of the leaf. *Nilsonia tenuinervis* Seward is a common fossil at Hasty
Bank where it is sometimes associated with *Nilssoniopteris vittata*
(Brongniart) Harris, although the latter is rare. Interestingly, in the
Gristhorpe Plant Bed it is *N. vittata* which is very common and *N.
tenuinervis* which is rare. At Scalby Ness only *N. tenuinervis* has been
found so far, which makes determination easy.

  Nilsonia compta (Phillips) Bronn (Pl. 9, fig. 1) is one of the most
common fossils in the Gristhorpe Plant Bed. It is found on only rare
occasions at Hayburn Wyke. Leaves are typically 500 mm long but usually
fragmentary, and 40 mm wide. The lamina is divided into segments that
are often very uneven in width; commonly broader than long in the lower

---

EXPLANATION OF PLATE 9

Fig. 1. *Nilsonia compta* (Phillips) Bronn; S. 8523; × 1.
Fig. 2. *Nilsonia kendallii* Harris; S. 4775; × 1.
Figs 3–4. *Beania gracilis* Carruthers; × 2. 3, S. 2309; stalk with part of a seed
  attached (see arrow); a *Nilsonia compta* leaf occurs in the lower left-hand
  corner. 4, S. 1245; seed.
Fig. 5. *Paracycas cteis* Harris; S. 1477; × 1.
Fig. 6. *Androstrobus prisma* Thomas and Harris; S. 1410; two microsporophylls; ×
  1.
Figs 1, 3–4 from the Gristhorpe Plant Bed; figs 2, 5–6 from Hasty Bank.

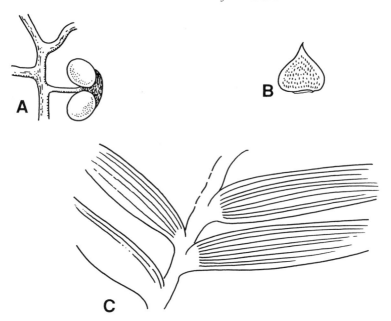

TEXT-FIG. 27. Cycads. A, *Beania gracilis* Carruthers; axis with one megasporophyll bearing two seeds. B, *Deltolepis crepidota* Harris; scale leaf. C, *Pseudoctenis oleosa* Harris; after Harris (1964). All × 0·7.

part of the leaf, then becoming equal, and longer than broad in the middle and upper part. The shape of the segments can vary between rhomboidal with lateral margins curving forwards; rhomboidal with strictly parallel margins; and an almost triangular shape.

Occasionally in the Gristhorpe Plant Bed, one can find the male fructification of *Nilsonia compta* (Phillips) Bronn known as *Androstrobus manis* Harris (Pl. 8, fig. 4). These cones resemble living male cycad cones and are approximately 20 mm wide and at least 50 mm long, containing many spirally arranged microsporophylls covered with pollen sacs on their lower side. On first view, these male cones can be mistaken for female

---

EXPLANATION OF PLATE 10

Fig. 1. *Nilsonia syllis* Harris; S. 1383; Hasty Bank; × 1.
Fig. 2. *Nilsonia tenuicaulis* (Phillips) Fox-Strangways; S.1412; note minute apical teeth (arrow); Gristhorpe Plant Bed; × 2.

conifer cones such as those of living *Abies* or *Picea*, but of course they bear no relation whatsoever.

More common are isolated seeds of the female fructification *Beania gracilis* Carruthers (Pl. 9, fig. 4). Complete cones (at least 100 mm long, 60 mm wide, bearing megasporophylls in a very loose spiral) are extremely rare, but detached megasporophylls are locally common (Pl. 9, fig. 3). They consist of a longitudinally striated, approximately 20 mm long stalk and a broadly rhomboidal or oval head (usually 18 by 6 mm) carrying one pair of fleshy seeds (around 10 mm in diameter) (Text-fig. 27A).

Deltolepis crepidota Harris (Text-fig. 27B) is the scale leaf belonging to the same plant. These scale leaves occur on living cycad stems and it is generally presumed that this is also the case for their fossil counterparts. At the base, the scales are about 15 mm wide and 18 mm long, including the acute apex. The surface is thickly marked with short longitudinal ridges.

The fact that the female fructifications are rather different from those of living cycads (megasporophylls are scale-like instead of leaf-like), has led palaeobotanists to assign *Nilsonia* to a different cycadalean family, Nilsoniaceae.

Nilsonia kendallii Harris (Pl. 9, fig. 2) is found in abundance at Hasty Bank. It has much shorter leaves, rarely exceeding 150 mm and often less than 100 mm. The width is 10–40 mm (typically 20 mm). The lamina segments in the middle region of the leaf are variable in shape but mainly triangular, about 10 mm long and 8 mm broad at their base; the segments can also be rectangular and rhomboidal in shape. The segments usually point slightly forwards. The veins are fine and typically 35 occur per 10 mm^2 near the base of the segments.

Nilsonia syllis Harris (Pl. 10, fig. 1) can be found at Hasty Bank and at Hayburn Wyke, but at both localities it is rather rare. It is a large leaf and may exceed 1 m long; its width in the middle and upper parts is about 100 mm. The lamina segments typically curve forwards and are about 50 mm long, 4–6 mm wide at their base, and have an acute apex. The veins are fairly distinct, about 20 per 10 mm near the rachis.

Nilsonia tenuicaulis (Phillips) Fox-Strangeways (Pl. 10, fig. 2) is known only from the Gristhorpe Plant Bed where it is locally common.

EXPLANATION OF PLATE 11

Figs 1, 4. *Pseudoctenis lanei* Thomas; S. 2699; Hasty Bank. 1, × 1. 4, detail of fig. 1, showing venation; × 2.

Figs 2–3. *Ctenis kaneharai* Yokoyama; S. 1416; Hasty Bank. 2, × 1. 3, detail of fig. 2, showing venation on the right hand side; × 2.

The leaf is 200–300 mm long and typically 50–80 mm wide. The lamina segments in the middle part of the leaf never curve forward; they are 40 mm by 10 mm basally, quickly contracting to a width of about 6 mm and then only slightly diminishing to about 4 mm near their obtuse or truncate apex, sometimes with minute teeth. The veins are fine, with a concentration of 10–20 per 10 mm near the base.

Paracycas

Paracycas cteis Harris (Pl. 9, fig. 5) is an easily recognizable fossil at Hasty Bank; it has long, narrow leaf segments. The shape of the leaf as a whole is unknown, its width is at least 150 mm, its length 200 mm but probably much longer. The pinnae arise at almost right angles to the rachis and taper very gradually from 1·5–2·5 mm at their base to two-thirds of this width at around 50–70 mm from the rachis. The nature of their apex is unknown. Pinnae are single-veined.

Pseudoctenis

Although more species of this genus are known from Yorkshire, only two are dealt with here as the others are very rare. What is more, apart from these two species, no other members of the genus have been collected from the plant beds discussed here. *P. lanei* Thomas (Pl. 11, figs 1, 4) is common at Hasty Bank, although it can also be found in the Gristhorpe Plant Bed. The pinnae are narrower than in the following species (6–10 mm broad in the lower part of a leaf, 4–6 mm in the upper part). The pinna base tapers distinctly, and is usually decurrent. Adjoining pinnae are typically separated by a distance which is about equal to their own width. Veins are parallel, forking near the pinna base, with a concentration in the middle region of around 20 per 10 mm. At Hasty Bank, *P. lanei* Thomas can be accompanied by its male fructification *Androstrobus prisma* Thomas and Harris. Complete cones are very rare, but detached microsporophylls are locally common (see Pl. 9, fig. 6). These are more-or-less rhomboidal in outline, about 18 mm wide and 10 mm high. The outermost (upper) 5 mm are sterile but the inner parts are thickly covered with

---

EXPLANATION OF PLATE 12

Fig. 1. *Nilssoniopteris vittata* (Brongniart) Harris; S. 1348; × 1.

Fig. 2. *Ctenis sulcicaulis* (Phillips) Ward; S. 1223; × 0.75.

Figs 3–4. *Williamsoniella coronata* Thomas. 3, S. 1349; complete flower associated with leaf fragment of *Nilssoniopteris vittata*; × 1. 4, part of receptaculum with seeds, interseminal scales and corona; photo courtesy of Dr C. R. Hill; × 2·5.

All specimens from the Gristhorpe Plant Bed.

microsporangia (pollen sacs), about 0·5 mm wide and up to 1·5 mm long. The shape of the microsporangia is roughly like a prism (hence the species' name), usually with four to six angles but sometimes rounded.

Pseudoctenis oleosa Harris (Text-fig. 27c) also occurs at Hasty Bank, but more rarely than *P. lanei*. The former is distinguished by its more or less contracted pinna base, and larger pinnae with only a small distance between adjoining pinnae. Typical pinnae are 15 mm wide (rarely up to 30 mm) and may be 120–200 mm long. Near the apex of a leaf the pinnae are smaller (6 mm by 100 mm). The veins are often inconspicuous, parallel and often fork in the basal part of the pinna. Their concentration is typically 15 veins per 10 mm.

Ctenis

These species can easily be distinguished from other cycads by their anastomosing venation. The Yorkshire species discussed here each occur at only one of the localities: *C. kaneharai* Yokoyama is locally common at Hasty Bank and *C. sulcicaulis* (Phillips) Ward in the Gristhorpe Plant Bed. *C. sulcicaulis* (Pl. 12, fig. 2) is a large leaf, perhaps up to 700 mm long and 250 mm wide. Usually only pinna fragments are found, but in instances where the rachis is preserved expanded pinna bases with decurrent lower pinna margins can be clearly seen. The pinnae are 100–150 mm long and around 10 mm wide. The veins are nearly parallel, anastomosing at intervals of 10–20 mm, with typically around 15 veins per 10 mm.

The leaf of *Ctenis kaneharai* Yokoyama (Pl. 11, figs 2–3) is also large. Although the total length of the leaf is unknown, with a known width of 300–400 mm, *C. kaneharai* may be even larger than *C. sulcicaulis*. The rachis is about 10 mm wide and longitudinally striated, bearing pinnae with contracted bases. These pinnae are 150–200 mm long and 10–15 mm wide. The veins are nearly parallel, anastomose at 10–30 mm intervals and have a concentration of 10–15 per 10 mm near the base.

11. BENNETTITALES

This is an order of plants which occurred globally in the Mesozoic but became extinct in the Cretaceous. As mentioned above, although some bennettitalean leaves closely resemble those of the cycads, they can be distinguished on the basis of cuticle character and venation pattern (secondary veins are usually unforked in cycads, whereas they are once forked in bennettitalean leaves). Leaf forms which typify the genera *Zamites, Otozamites* and *Ptilophyllum* are unknown in the cycads.

Key to the bennettitalean foliage (compare Text-fig. 28)

1. Leaf simple 2
   Leaf divided into pinnae 3
2. Leaf length typically four times width; leaves usually 30–50 mm wide
   ....*Nilssoniopteris major*
   Leaf length typically ten times width; leaves usually 15–25 mm wide
   ....*Nilssoniopteris vittata*
3. Pinnae not contracted basally 4
   Pinnae contracted basally (at least on one side)............................5
4. Pinnae about as long as broad ....*Anomozamites nilssonii*
   Pinnae at least twice as long as broad (up to ten times) ....
   ....*Pterophyllum thomasii*
5. Lower basal angle of pinna expanded...6
   Lower basal angle of pinna contracted 8
6. Pinnae usually less than 10 mm long ....*Ptilophyllum pecten*
   Pinnae usually more than 10 mm long 7
7. Pinnae up to 3 mm wide ....*Ptilophyllum pectinoides*
   Pinnae more than 4 mm wide ....*Ptilophyllum hirsutum*
8. Pinna base symmetrical (equally contracted on both sides)..................
   ....*Zamites gigas*
   Pinna base asymmetrical; the upper basal angle enlarged as an auricle
  9
9. Leaf more than 40 mm wide..10
   Leaf less than 40 mm wide; pinnae small, nearly circular...................
   ....*Otozamites tenuatus*
10. Pinna length more than three times width...11
    Pinna length less than three times width; pinnae more or less ovate.....
    ....*Otozamites beanii*
11. Pinnae more than 4 mm wide just above auricle.*Otozamites graphicus*
    Pinnae less than 4 mm wide just above auricle ..*Otozamites gramineus*

Reproductive organs

The Yorkshire Bennettitales probably belong to the Williamsoniaceae, a family of large plants (*c.* 2 m tall) characterized by stalked cones or 'flowers'. These flowers have one or more whorls of bracts (belonging to the genus *Cycadolepis*) surrounding the base. Leaf scars indicate that the leaves were borne in a crown at the end of a stem (*Bucklandia*). Most members of the Williamsoniaceae have separate male and female flowers, known as *Weltrichia* and *Williamsonia* respectively. In some instances, examples have been found where the central female part of the flower is surrounded by a whorl of wedge-shaped microsporophylls (Pl. 12, fig. 3; Text-fig. 29). Such specimens are thought to be bisexual flowers and are known as *Williamsoniella*.

Several species of *Williamsonia* occur in Yorkshire. In general, these flowers are rounded and consist of a dome-shaped receptacle whose lower part bears numerous seeds and interseminal scales (Text-fig. 29). The term corona refers to the uppermost part of the receptacle which did not bear seeds (Pl. 12, fig. 4). Micropyles of the seeds project beyond the interseminal scales. Bracts surrounded the flowers and probably offered protection, until they fell off at a later stage. In contrast, the male 'flower' *Weltrichia* is characterized by a massive cup which in its outer part is divided into a number of equal lobes which taper to a point (see Text-fig. 29). The inner side of the cup and the lobes bear rows of pollen sacs, either directly, or on appendages. Where relevant, fructifications will be discussed in relation to the leaf species to which they are attributed.

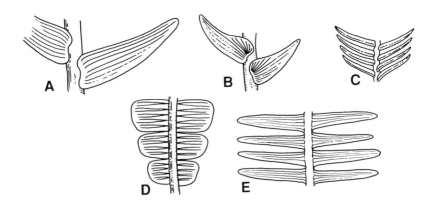

TEXT-FIG. 28. Schematic drawing of the leaf shape and venation of some bennettitalean leaves. A, *Zamites*. B, *Otozamites*. C, *Ptilophyllum*. D, *Anomozamites*. E, *Pterophyllum*.

Nilssoniopteris
This is the only genus with an undivided lamina. If it were not for differences in venation pattern, *N. vittata* (Brongniart) Harris could easily be confused with the cycad species *Nilsonia tenuinervis* Seward; unlike *Nilsonia*, *Nilssoniopteris* has secondary veins which tend to be forked once. Although *Nilssoniopteris major* (Lindley and Hutton) Harris (Pl. 13, fig. 1) is a rare fossil in the Gristhorpe Plant Bed, full-sized leaves have been found which are some 150 mm long and 30–50 mm wide, with a short petiole, and a thin lamina revealing once forked secondary veins. *N. vittata* (Pl. 12, fig. 1) is one of the most common fossils in the Gristhorpe Plant Bed. Here, a full-sized leaf is typically 250 mm long and 15–25 mm

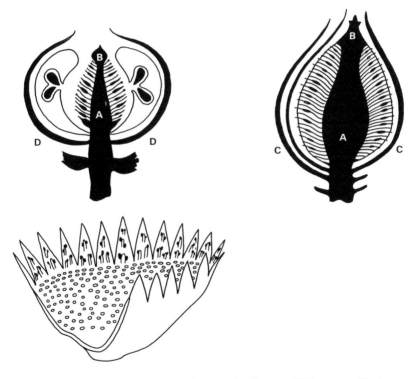

TEXT-FIG. 29. Schematic drawings of bennettite flowers. *Williamsoniella* (upper left), flower with A, the receptaculum bearing ovules and interseminal scales, B, the corona, C, the microsporophylls, and D, the surrounding bracts. *Williamsonia* flower (upper right) with A, the receptaculum bearing ovules and interseminal scales, B, the corona, and c, the surrounding bracts. *Weltrichia* flower (bottom left). After Harris (1969).

TEXT-FIG. 30. *Cycadolepis stenopous* Harris; scale leaf, after Harris (1969); × 3.

wide. Petioles are short and stout, and the substance of the lamina thick. The veins are simple or forked once, and not as pronounced as in *N. major.*

Detached megasporophylls from the bisexual flower *Williamsoniella coronata* Thomas, belonging to *Nilssoniopteris vittata* (Brongniart) Harris, can also be found in the Gristhorpe Plant Bed. These microsporophylls are generally 10–15 mm long and carry four pollen sacs. Whole flowers (*c.* 20 mm long) are found rarely (Pl. 12, fig. 3; Text-fig. 29), but separate female parts (gynoecium), from which bracts and microsporophylls have disarticulated, occur more commonly (Pl. 12, fig. 4). The lower portion of the gynoecium is typically 10 mm wide, and the receptacle is covered with seeds and interseminal scales in its lower and middle parts. This is also the case for *Williamsonia,* where the upper 2–3 mm of receptacle is bare and called the corona. The bracts around the *Williamsoniella* flower have not received a name of their own, as they are almost always found attached.

Anomozamites *and* Pterophyllum
Both of these genera are distinguished from all other bennettitalean leaves by their straight, at times slightly expanded, pinna base. The two genera are distinguished on the basis of simple leaf morphology: those pinnae (or leaf segments) which are roughly equidimensional are ascribed to the genus *Anomozamites*, while pinnae which are more than twice as long as broad are referred to *Pterophyllum*. One species of each genus is described below.

EXPLANATION OF PLATE 13

Fig. 1. *Nilssoniopteris major* (Lindley and Hutton) Harris; S. 1521, two full-sized, almost complete leaves; × 0·5.
Fig. 2. *Anomozamites nilssonii* (Phillips) Harris; S. 3085; × 1.
Fig. 3. *Pterophyllum thomasii* Harris; S. 1322; × 1.
All specimens from Cloughton Wyke, near Scarborough.

Anomozamites nilssonii (Phillips) Harris (Pl. 13, fig. 2) is known from the Gristhorpe Plant Bed where it is locally common. Leaves are typically 150 mm by 30 mm, and the lamina divided into segments which may be square or uneven. The free margin of these segments is minutely dentate, marking the end of a vein, which are parallel and usually once forked. Occasionally *A. nilssonii* is accompanied by tiny scale leaves known as *Cycadolepis stenopus* Harris. These, which are by far the smallest of *Cycadolepis* species, are about 8 mm long and 4 mm wide, lanceolate, with an acute apex and a constricted base (see Text-fig. 30).

Pterophyllum
Although rare, *Pterophyllum thomasii* Harris (Pl. 13, fig. 3) can be found in both the Gristhorpe Plant Bed and at Hayburn Wyke. The total length of the leaf is unknown, but considering that the typical width is 100–150 mm it must have been large in size. The rachis is up to 8 mm wide, with strong petioles at least 50 mm long. The pinnae are variable in size (30–100 mm long and 2–4 mm wide) and arise on the side of the rachis at an angle of 80–90° before curving slightly forwards. The apices are acute.

Otozamites
The genera *Zamites, Otozamites* and *Ptilophyllum* can be distinguished by the base of their pinnae alone. The pinnae of *Otozamites* are characterized by having an enlarged upper basal angle (auricle); in *Zamites* the pinnae have a constricted base on both sides; and the pinnae have a slightly decurrent lower basal angle in *Ptilophyllum*. For most species it is easy to ascertain to which genus a specimen belongs. However, the characters of some species are rather less distinct, which can present difficulties in their identification.

Otozamites graphicus (Leckenby) Phillips (Pl. 14, fig. 2) is a typical *Otozamites* species in that the pinnae have a distinct auricle. It can be found at all four localities discussed here, rarely at Hasty Bank, Hayburn Wyke and in the Gristhorpe Plant Bed, but commonly at Scalby Ness. *O. graphicus* must have been large in size, as the width of the middle region of a full-sized leaf is about 100 mm. This gives an estimated length of 500–700 mm. Pinnae are typically 50 mm long and 8 mm wide above the

---

EXPLANATION OF PLATE 14

Fig. 1. *Otozamites gramineus* (Phillips) Phillips; S. 6432; Hayburn Wyke; × 1.

Fig. 2. *Otozamites graphicus* (Leckenby) Phillips; S. 1338; Scalby Ness; × 1.

Fig. 3. *Otozamites beanii* (Lindley and Hutton) Brongniart; S. 3011; Gristhorpe Plant Bed; × 1.

Fig. 4. *Ptilophyllum hirsutum* Thomas and Bancroft; S. 3032; Scalby Ness; × 0·5.

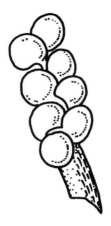

TEXT-FIG. 31. *Otozamites tenuatus* (Leckenby) Phillips; characteristic small, almost circular leaflets, after Harris (1969); × 1.

auricle, becoming shorter and narrower in the upper part where the auricle is less obvious. Near the leaf base pinnae are shorter, but still broad and with a clear auricle. Pinna apices are usually acute.

Otozamites gramineus (Phillips) Phillips (Pl. 14, fig. 1) is restricted to the Saltwick Formation, at the Hasty Bank and Hayburn Wyke localities where it is rare. Good specimens can be found, nonetheless. The leaf is of moderate length (*c.* 300 mm) and width (50–70 mm), composed of slender pinnae up to 35 mm long and 3 mm wide just above the distinct auricle. Shorter pinnae tend to occur near the apex and base of the leaves. The apices of the pinnae themselves are acute.

Hayburn Wyke is the main locality for *Otozamites tenuatus* (Leckenby) Phillips, but finds are rare. It is at once distinguished by small, almost circular pinnae between 3 mm and 6 mm wide, with obtuse apices and truncate or slightly cordate bases. The auricle is only faintly visible. The rachis is completely covered by the pinnae and the leaves are as a whole linear, up to 300 mm long and only 6–10 mm wide (see Text-fig. 31).

EXPLANATION OF PLATE 15

Fig. 1. *Zamites gigas* (Lindley and Hutton) Morris; S. 1319; Hayburn Wyke; × 0·5.

Fig. 2. *Weltrichia sol* Harris; S. 1344; note cup (arrow A) and spreading microsporophylls (arrows B); upper left hand corner shows part of *Zamites gigas* leaf; Hayburn Wyke; × 0·5.

1

2

Otozamites beanii (Lindley and Hutton) Brongniart (Pl. 14, fig. 3) resembles *O. tenuatus* in the general shape of the pinnae, but otherwise tends to be ovate and much larger (typically 30 by 15 mm, but more equidimensional and rounded near the leaf base). The auricle is also clearer. The leaf as a whole is wider than *O. tenuatus* (*c.* 50 mm, in contrast with 6–10 mm) and up to 300 mm long. *O. beanii* is a rare constituent of the Gristhorpe Plant Bed.

Zamites

Zamites gigas (Lindley and Hutton) Morris (Pl. 15, fig. 1) is a common fossil at Hayburn Wyke, where beautiful leaf fragments can be collected. Although the pinnae are quite large (50–80 mm long and 6–10 mm wide), the leaf itself is only of moderate size, typically 300 mm by 120 mm. The pinnae bases are symmetrically contracted and the lower two-thirds parallel-sided, tapering in the upper third towards an acute apex.

At Hayburn Wyke, *Z. gigas* is often accompanied by male and female flowers, known as *Weltrichia sol* Harris and *Williamsonia gigas* Carruthers respectively. *W. gigas* (the type species of *Williamsonia*) is the largest female flower of the genus found in Yorkshire. The flower is persistent on the peduncle, and has a receptacle which is about 50–60 mm long and 20–30 mm wide at the base, tapering to 10 mm wide above (Text-fig. 29B). The receptacle then expands slightly into the conically shaped corona which is about 15 mm long. Seeds and interseminal scales cover the receptacle and are normally shed at maturity, with the exception of those occurring just below the corona. So-called 'perianth-scales' occur persistently around the gynoecium. These robust scales are numerous, usually 100 mm by 13 mm in size, and because they are always found attached to the gynoecium have not received a name of their own. Instead they are related by shape to *Cycadolepis* scales known from other bennettitalean flowers; only the perianth-scales of *Williamsonia gigas* Carruthers are larger. Usually they enclose the whole gynoecium, although in a mature stage they can be recurved.

EXPLANATION OF PLATE 16

Fig. 1. *Ptilophyllum pecten* (Phillips) Morris; S. 14952; × 1.

Fig. 2. *Williamsonia leckenbyi* Nathorst; S. 2338; part of a gynoecium (between arrows); × 1.

Fig. 3. *Cycadolepis nitens* Harris; S. 2680; × 2.

Fig. 4. *Ptilophyllum pectinoides* (Phillips) Halle and *Cycadolepis hypene* Harris; S. 2700; × 1.

Figs 1–3 from the Gristhorpe Plant Bed; fig. 4 from Hasty Bank.

Weltrichia sol Harris (Pl. 15, fig. 2; Text-fig. 29C) is a large male flower composed of an open cup, some 100 mm wide at the top and 70–80 mm high. The Rays are numerous (usually up to 30), 10 mm wide at their base and tapering to an acute apex, and 50–60 mm long; a complete flower might therefore be as large as 120–140 mm! The outer surface of the cup and rays often shows longitudinal ridges and transverse wrinkles, and the inner surface bears a large number of resin bodies around 1 mm in diameter. Pollen sacs are about 2 mm long and 0·3 mm wide, and are found in two rows on short appendages originating from the inner surface of the rays.

Ptilophyllum
The three *Ptilophyllum* species are also distinct in terms of their occurrence: *P. pectinoides* (Phillips) Halle occurs at Hasty Bank, *P. pecten* (Phillips) Morris in the Gristhorpe Plant Bed and *P. hirsutum* Thomas and Bancroft at Scalby Ness.

Ptilophyllum pecten (Phillips) Morris (Pl. 16, fig. 1) is locally abundant in the Gristhorpe Plant Bed, but in certain places is completely absent. A complete leaf is usually about 200 mm long and 12 mm wide, with pinnae arising at an angle of 60° to the rachis. The pinnae tend to contact laterally, although gaps of 0·5 mm occur occasionally. On average, the pinnae are 1·0–1·5 mm wide, which is approximately one-fifth of their total length (up to 7 mm). Apices are obtuse. Veins are unclear due to the thickness of the lamina, but where visible they are almost parallel. In rare instances *P. pecten* is accompanied by its female flower, *Williamsonia leckenbyi* Nathorst, male flower *Weltrichia pecten* (Leckenby) Harris and scale leaf *Cycadolepis nitens*.

Unlike *W. gigas*, *Williamsonia leckenbyi* Nathorst (Pl. 16, fig. 2) has a gynoecium which is detached from the peduncle and surrounded by bracts (or scale-leaves which are almost always found detached). The lower part of the gynoecium is formed by a layer of coherent interseminal scales which are usually preserved along with seeds. The corona has the shape of a truncated cone, 2 mm long and 2·5 mm wide at the base. *Weltrichia pecten* (Leckenby) Harris is a detached flower consisting of a wide cup (30–50 mm wide) which divides into ten to 12 rays, which are 30 mm long

---

and curve slightly inwards at the apex. The cup has longitudinal striations, but no transverse wrinkles or hairs. The rays bear two rows of short stalked, oval pollen sacs (3 mm by 1 mm) on their inner surface. *Cycadolepis nitens* (Pl. 16, fig. 3) is a small lanceolate scale leaf 12–15 mm long and 5 mm wide just above its base, with a mucronate apex. The substance of the leaf is very thick and glossy, and densely covered with short fine hairs.

Ptilophyllum pectinoides (Phillips) Halle (Pl. 16, fig. 4) is abundant at Hasty Bank. The leaves are about 200 mm long and 20–40 mm wide, the pinnae straight and arising at an angle of 60° to the rachis. Each is separated by a gap of 0·5 mm; the pinnae are rarely found in contact. The pinnae are 1·8 mm wide, and constitute about one-tenth of the length (15–20 mm, occasionally up to 40 mm). Apices are often acute, and the substance of lamina rather thick, usually concealing unbranched veins.

Hasty Bank specimens of *P. pectinoides* are infrequently found with the related female and male flowers *Williamsonia hildae* Harris and *Weltrichia whitbiensis* (Nathorst) Harris respectively, and more commonly found with the scale leaf *Cycadolepis hypene* Harris. *W. hildae* (Pl. 17, figs 2–3) resembles *W. leckenbyi* Nathorst quite closely, differing only in having a slightly larger corona. This feature measures 4 mm wide at the base, narrowing slightly to 3 mm, and typically reaches only 2 mm high. *Weltrichia whitbiensis* (Pl. 17, fig. 1) is composed of a wide cup which is usually 40 mm across, 25 mm high, and divided into around 15 rays nearly 40 mm long. The outer surface of the cup and rays is longitudinally striated, but transverse wrinkles and hairs are absent. On the inner surface of the rays, two rows of oval pollen sacs are found. *Cycadolepis hypene* (Pl. 16, fig. 4) is a lanceolate scale leaf with a filiform apex; it is about 25 mm long and 6 mm wide just above the base. The substance is glossy and thick, with longitudinal fibres projecting on the surface as ridges. The margins are often covered with short, fine hairs.

Ptilophyllum hirsutum Thomas and Bancroft (Pl. 14, fig. 4) is the largest and least typical of the *Ptilophyllum* species. The leaf is 40 mm wide in the middle part and is estimated to be 200–400 mm long. The lower basal angle is more nearly straight than decurrent; pinnae arise at 70–80°. These are 5–7 mm wide, 1·5–2·5 mm long, and nearly straight with an obliquely truncate apex. Veins are conspicuous, nearly parallel and branch occasionally. The lower surface of the leaf is covered with thick papillae which can be seen with a magnifying glass or hand lens. *P. hirsutum* is a rare fossil at Scalby Ness.

12. GINKGOALES

Ginkgoales is an ancient group which first appeared in the Late Palaeozoic. Today, they are represented by the deciduous maidenhair tree *Ginkgo biloba* Linnaeus alone; a temple tree native to China and Japan, but now cultivated in many temperate areas where it has been artificially introduced. Its frost-resistant nature, ability to withstand air pollution, and decorative character make *G. biloba* a popular tree.

One distinctive characteristic of *G. biloba* is the stalked petiolate, fan-shaped leaf with almost parallel, slightly diverging, dichotomously branching veins. The leaves are clustered at the end of short shoots, which are in turn borne on longer shoots. These are shed singly, in contrast with the Czekanowskiales, in which the leaves remain attached in groups to the short shoot and the unit is shed in its entirety. Even within one tree, there is a considerable amount of variability in leaf shape, ranging from almost entire fan-shaped leaves to deeply notched varieties. Male and female (ovulate) structures are borne on separate plants. The male structures resemble catkins and are borne in the axils of bud scales or leaves on the short shoots. The microsporophylls bear two to four pollen sacs. Female structures are also borne on short shoots and consist of a stalk that terminates in two or three ovules which after fertilization ripen to form seeds.

During the Mesozoic, the Ginkgoales were far more diverse and widespread than today, but tended to be restricted to areas with a temperate climate. Ginkgoales, together with Czekanowskiales, are thus

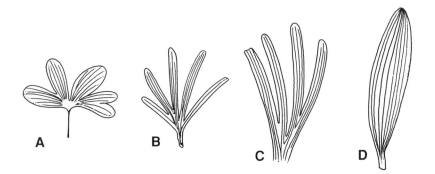

TEXT-FIG. 32. Schematic drawing of the characteristic leaf shape and venation of some ginkgoalean leaf genera. A, *Ginkgo*. B, *Baiera*. C, *Sphenobaiera*. D, *Eretmophyllum*.

used as indicators of cooler climates and it is interesting to note that there is evidence that fossil Ginkgoales were deciduous. Most Mesozoic Ginkgoales are attributed to the group purely on the basis of leaf-shape. Indeed, for those Jurassic *Ginkgo*-species with leaves resembling those of the extant maidenhair tree, the generic name *Ginkgo* is given. As is the case of living *G. biloba*, the variability of leaf shape within the fossil species is great, particularly in the deepness of lamina divisions.

The few Mesozoic ginkgoalean fructifications that have been recovered resemble those of *G. biloba* rather closely (Zhou 1991). Similarly, the male catkin and seeds of one of the Yorkshire *Ginkgo* species show many related characteristics to extant material (van Konijnenburg-van Cittert 1971; Harris *et al.* 1974).

Key to ginkgoalean foliage (see also Text-figure 32)

1. Lamina lanceolate, undivided 2
   Lamina wedge- or fan-shaped (usually lobed or divided) 3
2. Leaf 50–100 mm long, widening gradually from narrow base ....
   ....*Sphenobaiera gyron*
   Leaf up to 150 mm long and distinctly petiolate ......*Eretmophyllum* spp.
3. Leaf as a whole wedge-shaped; no distinction between lamina and
   petiole 4
   Leaf as a whole fan-shaped; petiole abruptly enlarging into lamina 5
4. Lamina bilobed ....*Sphenobaiera gyron*
   Lamina divided into many segments ....*Sphenobaiera pecten*
5. Lamina lobed or coarsely divided; more than four veins per segment...6
   Lamina finely divided; one to four veins per segment ......*Baiera furcata*
6. Lamina less than 25 mm in radius.............................*Ginkgo whitbiensis*
   Lamina more than 25 mm in radius...7
7. Basal angle of lamina less than 100°; segments long, narrow.................
   ....*Ginkgo longifolius*
   Basal angle of lamina more than 100°; segments rarely more than 40
   mm long, usually broad..8
8. Basal angle often more than 180°; lamina shallowly divided.................
   ....*Ginkgo digitata*
   Basal angle less than 180°; lamina deeply divided into segments.........9
9. Lamina substance thin, number of segments around ten.*Ginkgo sibirica*
   Lamina substance robust, number of segments around six ....
   ....*Ginkgo huttonii*

The division between *Ginkgo* and *Baiera* is quite arbitrary and rather vague: specimens are often recovered with a varying number of veins per segment (e.g. *G. longifolius* where the number of veins varies between four and six), making classification difficult. There are several *Ginkgo*

TEXT-FIG. 33. A, *Ginkgo huttonii* (Sternberg) Heer; S. 3037; Scalby Ness; × 1·5. B, *Ginkgo digitata* (Brongniart) Heer; S. 12632; detail of Plate 17 figure 4 showing lamina segments and venation; Gristhorpe Plant Bed; × 2. C, *Ginkgo whitbiensis* Harris; S. 1468; Roseberry Topping, near Hasty Bank; × 2.

species in Yorkshire, but only one *Baiera* species (*B. furcata* (Lindley and Hutton) Braun).

Ginkgo

G. huttonii (Sternberg) Heer (Text-fig. 33A) is the best known Yorkshire *Ginkgo*. It is the most common fossil at Scalby Ness, and found on rare occasions at Hayburn Wyke. Only one example of a male catkin has ever been found in Yorkshire (van Konijnenburg-van Cittert 1971), but seeds are more often encountered, particularly at Scalby Ness. The leaf is characterized by a slender petiole, a lamina radius of 30–40 mm, and a basal angle of around 120°. The lamina is usually deeply divided into six segments, with rounded apices which can be irregularly truncate or notched. The veins are inconspicuous, about 25 found per 10 mm² half-way up the lamina, branching only in the lower part of the segments. The seeds (Pl. 17, fig. 5) are attributed to *G. huttonii* on the basis of associated leaves, and close resemblance to extant *Ginkgo* seeds (Harris *et al.* 1974). They are round and usually slightly less than 10 mm in diameter.

Ginkgo digitata (Brongniart) Heer and *Ginkgo longifolius* (Phillips) Harris both occur at Gristhorpe; the former locally common, the latter rare. *G. digitata* (Pl. 17, fig. 4; Text-fig. 33B) has a lamina radius of about

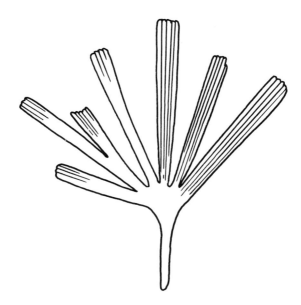

TEXT-FIG. 34. *Ginkgo longifolius* (Phillips) Harris; drawn after Harris *et al.* (1974); × 1.

30 mm and a basal angle of approximately 200°. The slender petiole carries a thin lamina with six to 12 segments and only shallow incisions. The apices are rounded or irregularly truncate. Veins are usually conspicuous, with around 16 per 10 mm occurring half-way up the lamina, branching throughout the whole lamina.

Ginkgo longifolius (Text-fig. 34) was originally known as *Baiera longifolia* (Phillips) Leckenby, until the affiliation was altered on the basis of four or more (five being typical) veins being found per segment (see Harris *et al.* 1974). The basal angle is about 90°, with a lamina radius of 60–65 mm. The lamina is deeply divided into four to eight segments which in their apical part may fork again. Each petiole is up to 2 mm wide and about 40 mm long. Because the leaf substance is thin, the veins are conspicuous. These are normally forked only once.

Ginkgo sibirica Heer and *Ginkgo whitbiensis* Harris are both rare fossils at Hasty Bank. The latter (Text-fig. 33C) is at once distinguished by its small size. The petiole is only 10–20 mm long, 2 mm wide, and the lamina narrow, with a radius of up to 20 mm and a basal angle around 90°. It is divided into two to four lobes with blunt, rounded apices. The veins are obscure.

Ginkgo sibirica Heer (Text-fig. 35A) has a large basal angle (100–200°, with a mean of 165°), a long petiole, and a maximum lamina radius of 40–50 mm. The lamina is usually divided into four to six main segments, each dividing again in their apical part (resulting in about ten segments). The apices are rounded. The leaf substance is fairly thin, the veins conspicuous with a concentration of around 15 per 10 mm half-way up the lamina (typically 7 veins per segment).

Baiera

Baiera furcata (Lindley and Hutton) Braun (Text-fig. 36A) is a rare fossil at Hayburn Wyke. It can be quite variable in both shape and size. The basal angle is 50–180° and the lamina radius is at least 70 mm. The petiole is long and slender, and the lamina is divided into four to 20 ultimate segments 1–4 mm wide. There are two to four inconspicuous veins per segment.

Eretmophyllum

This is the only genus in which the leaves tend to have a consistently undivided lamina. The petiole, about 20 mm long, widens rather abruptly into a sublanceolate lamina with a rounded apex. Veins are parallel and far apart, and fork only in the basal part of the lamina. There are two species of *Eretmophyllum* in Yorkshire which are almost indistinguishable on a macroscopic level (their cuticles are different): *Eretmophyllum pubescens* Thomas is locally common at Gristhorpe, and *Eretmophyllum whitbiense* Thomas is rare at Hasty Bank. *E. pubescens* (Text-fig. 35B) is a relatively

TEXT-FIG. 35. A, *Ginkgo sibirica* Heer; S. 1613; Roseberry Topping, near Hasty Bank. B, *Eretmophyllum pubescens* Thomas; S. 1262; Gristhorpe Plant Bed. C, *Sphenobaiera gyron* Harris; S. 7625; specimen almost indistinguishable from *Eretmophyllum*; Hasty Bank. All × 1.

TEXT-FIG. 36. A, *Baiera furcata* (Lindley and Hutton) Braun; S. 1259; Hayburn Wyke. B, *Eretmophyllum whitbiensis* Thomas; S. 2362; small leaf; Hasty Bank. C, *Solenites vimineus* (Phillips) Harris; S.14865; Cloughton Wyke. All × 1.

TEXT-FIG. 37. *Sphenobaiera pecten* Harris; characteristic shape of leaf, drawn after Harris *et al.* (1974); × 1.

large leaf which can be up to 150 mm long and 10–30 mm wide. The veins are conspicuous, far apart (5–10 per 10 mm), and converge slightly towards the apex. The leaf substance is rather thick.

E. whitbiense (Text-fig. 36B) is as long as *E. pubescens*, but usually somewhat narrower (8–25 mm wide). The leaf substance is of medium thickness, but the petiole and the margins near the lamina base are thicker. The veins are distinct and widely spaced (5–13 per 10 mm).

Sphenobaiera
This is distinguished from all other ginkgoalean leaves by the absence of a distinct petiole. The leaves as a whole are usually wedge-shaped. The lamina may be entire or forked into one or more times into segments. *Sphenobaiera gyron* Harris (Text-fig. 35C), which is common at Hasty Bank, has about half its leaves with an undivided lamina, the rest being bilobed. The undivided leaves can easily be confused with *Eretmophyllum whitbiense*. However, the lack of a petiole (the leaf widens gradually from

its narrow base), and the smaller size (50–100 mm long, and only 5–15 mm wide) of *S. gyron* ensure that the species remain distinct. The veins are inconspicuous, parallel and normally only fork once in the lower part of the lamina, whereas in the middle part of the leaf they have a concentration of 10–20 per 10 mm.

Sphenobaiera pecten Harris (Text-fig. 37) is rare at Hayburn Wyke. It is a wedge-shaped leaf, up to 100 mm long and 70 mm wide, and is divided by up to six successive dichotomies into a large number of narrow segments. Ultimate segments are often less than 1 mm wide. Veins are indistinct, with usually only one found in narrow segments, and two to four in the lower, broader segments. Superficially, *S. pecten* might be confused with *Baiera furcata* (Lindley and Hutton) Braun, but the latter has a petiole and fewer dichotomies, resulting in a much more 'open' leaf shape.

13. CZEKANOWSKIALES

This poorly understood group of gymnosperms was originally included in the Ginkgoales, because their leaves are borne on short shoots, which are shed in their entirety. Based primarily on the realization that the fructifications (especially the female ones) are rather different, Czekanowskiales is today regarded as a separate group of gymnosperms.

The leaves of Czekanowskiales are borne in bundles on short shoots surrounded by small scale leaves. The foliage leaves are elongated and simple (in the case of the genus *Solenites*) or dichotomizing (*Czekanowskia*). There is a single vein at the base of each leaf which normally forks several times, supplying each ultimate segment with a vein. Female fructifications belong to the genus *Leptostrobus*. They consist of an elongated axis bearing scale leaves below and loosely arranged seed-containing capsules above. The seed capsules are almost sessile, roundly wedge-shaped, and are composed of upper and lower valves of similar appearance (see Text-fig. 38). The valves have a slightly lobed terminal margin, and bear a row of small seeds on their inner surface.

The Yorkshire Jurassic flora contains the only known species of *Solenites, S. vimineus* (Phillips) Harris, and four *Czekanowskia* species. Moreover, the female fructification *Leptostrobus cancer* Harris can be

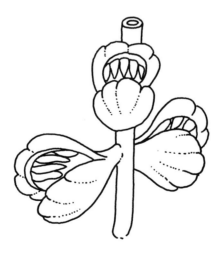

TEXT-FIG. 38. Restoration of part of a *Leptostrobus cancer* Harris cone, after Harris *et al.* (1974); × 3.

found in association with *Solenites vimineus*. Presumed male fructifications belong to the genus *Ixostrobus*. They are not discussed here, as they are extremely rare in Yorkshire.

Key to the foliage of the Czekanowskiales

1. Leaves entirely unbranched ....*Solenites*
   Leaves forked at least once 2
2. Leaves usually only forked once 3
   Leaves forked twice to six times...4
3. Leaves about 1 mm wide, over 120 mm long........*Czekanowskia furcula*
   Leaves considerably less than 1 mm wide, and around 80 mm long ....
   ....*Czekanowskia thomasii*
4. Leaves forked twice ....*Czekanowskia blackii*
   Leaves forked three times or more.................*Czekanowskia microphylla*

Solenites

Solenites vimineus (Phillips) Harris (Pl. 18, fig. 1; Text-figs 36C, 39A) is rare at Hasty Bank but quite common in the Gristhorpe Plant Bed, where it is found associated with *Leptostrobus cancer* Harris. The long foliage leaves are often found attached to small (5 mm by 5 mm) short shoots covered with scales. The leaves occur in bundles of ten to 15, and are typically 150–200 mm long (width up to 1 mm) and unbranched. The apex is more or less acute.

Leptostrobus cancer Harris (Pl. 18, figs 2, 5; Text-fig. 38) consists of a slender axis bearing capsules at intervals of 5 mm. Complete cones (Pl. 18, fig. 5) are rare. The capsules are usually detached and some 5 mm in diameter (Pl. 18, fig. 2), with rounded valves and some obtuse marginal lobes. There may be up to five seeds in each valve.

EXPLANATION OF PLATE 18

Fig. 1. *Solenites vimineus* (Phillips) Harris; S. 4798; short shoot with lower part of a bundle of leaves; Cloughton Wyke; × 2.

Figs 2, 5. *Leptostrobus cancer* Harris; Cloughton Wyke. 2, S. 6380; capsule associated with *Solenites vimineus*; × 2. 5, S. 4795; part of a cone (arrowed); × 1.

Fig. 3. *Czekanowskia thomasii* Harris; S. 2638; Gristhorpe Plant Bed; × 2.

Fig. 4. *Czekanowskia blackii* Harris; S. 1292; Scalby Ness; × 1.

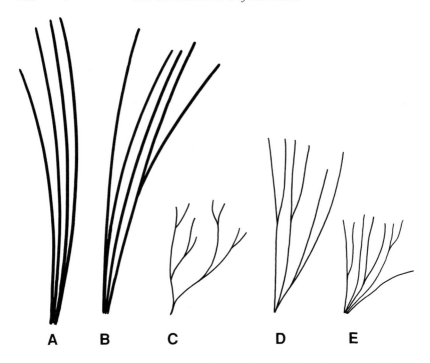

TEXT-FIG. 39. Schematic drawings of various czekanowskialean leaves. A, *Solenites vimineus* (Phillips) Harris. B, *Czekanowskia furcula* Harris. C, *Czekanowskia microphylla* (Phillips) Seward. D, *Czekanowskia blackii* Harris. E, *Czekanowskia thomasii* Harris.

Czekanowskia

Czekanowskia furcula Harris (Text-fig. 39B) can occasionally be found at Hayburn Wyke and, with the exception of the cuticle, resembles *Solenites vimineus* (Phillips) Harris rather closely. The leaves of each species are of about the same length, although the leaves of *C. furcula* are slightly wider (maximum of 1·2 mm) and fork once, usually in the middle area. The apex is acute.

Czekanowskia thomasii Harris (Pl. 18, fig. 3; Text-fig. 39E) is locally common at Gristhorpe. It can at once be recognized by its narrow foliage leaves (usually only 0·5 mm wide, but under a dichotomy maximum can reach 0·8 mm), which fork once. Their apex is acute. The leaves are around 8 mm long, and occur in bundles of seven or eight on tiny shoots 2 mm by 2 mm.

Czekanowskia microphylla (Phillips) Seward (Text-fig. 42C) occurs rarely at Gristhorpe. The foliage leaves are in bundles of about five, are 60–100 mm long and fork usually four or five times. One segment of a dichotomy is often shorter than adjacent segments. Ultimate segments are 5–8 mm wide, with an acute apex. Lower parts of foliage can be up to 1·2 mm wide.

Czekanowskia blackii Harris (Pl. 18, fig. 4; Text-fig. 39D) is locally common at Scalby Ness. The short shoot is approximately 3 mm by 3 mm and bears a bundle of five to eight foliage leaves. Although small fragments tend only to be found, the leaves are estimated to have been up to 150 mm long, normally forking twice, 0·7–0·8 mm wide (up to 1·2 mm below a dichotomy) and gradually tapering in their distal part to an acute apex.

14. CONIFERALES

Conifers vary in size from small shrubs to larger trees, and are characterized by woody tissue forming the stems, twigs and often the female cones. Conifers are today dominated by the evergreen variety, with fewer deciduous taxa; this also applies to fossil conifers.

The leaves of conifers are typically in the shape of needles or scales with only one vein, although some occur which are wider and contain several veins (e.g. the living genus *Agathis* and the fossil genera *Podozamites* and *Lindleycladus*). Male and female cones are normally borne on the same plant, more occasionally on different plants. Male cones consist of spirally arranged microsporophylls bearing a number of pollen sacs, whereas the woody female cones contain spirally arranged bract-scale complexes: a bract with, in its axil, an often partly fused female cone-scale.

Fossil coniferous leaves are grouped in so-called form-genera, based on scale- or needle morphology alone. A form-genus may therefore contain leaves that belong to different families. Some *Brachyphyllum* species from the Yorkshire Jurassic flora are classified in this way: for example, based on cone morphology, *B. mamillare* Lindley and Hutton belongs to the Araucariaceae and *B. crucis* Harris to the Cheirolepidiaceae. However, when fructifications (preferably female cones) are known, the conifers can be classified into more natural taxa and attributed to living or fossil families.

Sterile foliage forms a significant proportion of conifer finds in Yorkshire. Such fossils are difficult to identify in the absence of associated cones and are presently tentatively classified at a family level based on their similarities to known taxa. Fortunately, there are some Yorkshire conifers which are more fully understood due to the presence of male and/or female cones. Such species can be placed with more certainty in living or fossil families.

To date, the following families have been found in the Yorkshire Jurassic flora: Taxaceae (*Marskea jurassica* (Florin) Harris with male and female cones), Araucariaceae (*Brachyphyllum mamillare* Lindley and Hutton, male and female cones), Cheirolepidiaceae (*B. crucis* Harris, male and female cones), Podocarpaceae (*Cyparissidium blackii* (Harris) Harris, male and female cones), Pinaceae (*Pityocladus scarburgensis* Harris, female cones only) and possibly Podozamitaceae (*Lindleycladus lanceolatus* (Lindley and Hutton) Harris; with female cone scales, see Harris 1979; it is possible, however, that this plant may belong to another fossil family as different female cone scales are known from some *Podozamites* species). Some of these cones are very rare and will not be discussed here, others occur more commonly and some can even be found attached to foliage. Foliage, of course, occurs much more abundantly.

Key to conifer foliage

1. Leaves contracted basally then expanded to form a blade; lamina with several parallel veins.....................................*Lindleycladus lanceolatus*
   Leaves scale- or needle-like, with only one vein................................2
2. Leaves more than 25 mm long..3
   Leaves usually less than 25 mm long 4
3. Leaves usually 30–60 mm long, 1·5–2 mm wide; leaf surface glossy...
   ....*Bilsdalea dura*
   Leaves usually 50–100 mm long, 0·6–1·0 mm wide; not glossy ....
   ....*Pityocladus scarburgensis*
4. Length of leaves at least five times their width..................................5
   Length of leaves under five times their width 11
5. Leaves flat in section..6
   Leaves round or rhomboidal in section (equally thick in vertical and horizontal directions) 9
6. Leaves widest at their base...*Elatocladus laxus*
   Leaves (slightly) contracted at their base 7
7. Leaves usually less than 5 mm long ....*Elatocladus setosus*
   Leaves more than 5 mm long..8
8. Leaves more than 1·5 mm wide ....*Marskea jurassica*
   Leaves less than 1·5 mm wide.........................*Elatocladus zamioides*
9. Free parts of leaves stiff, straight or only very slightly falcate.............
   ....*Geinitzia rigida*
   Free parts of leaves typically falcate (curved forwards)....................10
10. Free leaf 6–12 mm long, falcate (curving forward) ....
    ....*Elatides williamsonii*
    Free leaf up to 5 mm long, strongly falcate................*Elatides thomasii*
11. Free parts of leaves shorter than (or equal to) width of its basal cushion 12
    Free parts of leaves longer than width of its basal cushion................13
12. Leaf and cushion (=part of leaf attached to the shoot) as long as broad
    ....*Brachyphyllum mamillare*
    Leaf and cushion twice as long as broad.............*Brachyphyllum crucis*
13. Free parts of leaves appressed to the stem ....*Cyparissidium blackii*
    Free parts of leaves diverging from the stem....................................14
14. Free parts of leaves thin, flat, lying in a horizontal plane ....
    ....*Elatocladus* spp.
    Free parts of leaves spirally arranged ....*Pagiophyllum insigne*

Elatocladus-*type foliage*

Four species of this type of foliage are dealt with here: *Marskea jurassica* (Florin) Harris and three *Elatocladus* species.

From our localities, the foliage of *M. jurassica* (Text-fig. 40A) is found only at Hasty Bank, where it occurs as very rare examples of entire shoots or, a little more commonly, separated leaves. The leaves are straight and arise with a contracted base from the shoot at an angle of 45°. The leaf pairs can be densely to widely spaced. Leaves are typically 20 mm long and 2 mm wide at the broadest part, which occurs in the lower third of the leaf. The upper part of the leaf gradually tapers to an acute apex. The midrib is prominent, especially in the lower half of the leaf. Based on male and female fructifications, *M. jurassica* is placed in the extant family Taxaceae. As these fructifications are extremely rare, no further discussion is provided here.

Elatocladus laxus (Phillips) Harris (Pl. 19, fig. 1) is commonly found in the Gristhorpe Plant Bed. It resembles *Marskea jurassica* (Florin) Harris in some aspects such as the size of the leaves, but in others it differs: the leaves arise at usually 90° to the stem (apically the angle can be smaller); the complete lateral shoot was shed, so no single leaves occur; the leaves spread in a horizontal plane by bending and twisting; the leaves are widest at their base, slowly tapering to an acute or acuminate apex. The midrib is shown as a prominent ridge on the lower side of the leaf and is sunken on the upper side. Although *E. laxus* closely resembles *Marskea jurassica*, no cones have been found and so it cannot safely be assigned to the Taxaceae.

Elatocladus zamioides (Leckenby) Seward (Text-fig. 40B) (rare in the Gristhorpe Plant Bed) is distinguished from the two species mentioned above by its narrower leaves (typically only 1·2 mm wide; length usually

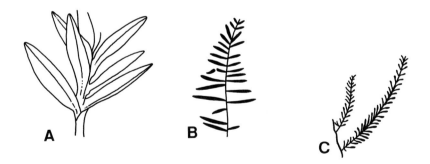

TEXT-FIG. 40. Conifer foliage, after Harris (1979). A, *Marskea jurassica* (Florin) Harris; × 2. B, *Elatocladus zamioides* (Leckenby) Harris; × 0·7. C, *Elatocladus setosus* (Phillips) Harris; × 0·7.

15–20 mm). The leaves are straight and are borne in opposite pairs at 80–90° to the stem (except near the apex where the angle is considerably smaller). Complete ultimate shoots were shed, so no single leaves occur. Leaf bases are typically contracted; the leaf blade is widest in the lower third, tapering towards an acute to acuminate apex. The midrib is sharply raised adaxially and prominent near the leaf base. Based on its unusual midrib and cuticular characteristics, *E. zamioides* may belong to the Cephalotaxaceae. Associated fructifications have yet to be found (Harris 1979).

Elatocladus setosus (Phillips) Harris (Text-fig. 40c) is a rare species found at Hayburn Wyke. The shoot systems are dropped, leaving their side branches and leaves intact. Leaves occur either in opposite pairs, or attached singly in a simple helix, and are borne from the main shoot at an angle of less than 30°, thus concealing the stem. On most lateral branches, leaves spread in a horizontal plane by bending and twisting their bases. Leaf blades diverge at 70–90°, are normally straight but near the apex are occasionally falcate and usually widely spaced (gaps wider than the leaves). Most leaves are only 5 mm long and 0·5–0·8 mm wide. The apex is acute to acuminate and the midrib inconspicuous. *E. setosus* has much shorter leaves than the other *Elatocladus* species described here and currently we know little of its taxonomic affinities.

Brachyphyllum

Brachyphyllum mamillare Lindley and Hutton (Pl. 19, fig. 2) is undoubtedly the most common conifer species found in Yorkshire and is one of the few species that occurs at all localities, if only as small shoot fragments. The leaves are usually small in size (typically only 1·5 mm long and 2·0 mm wide), although some larger forms can occasionally be found (up to 4 mm long and 3 mm wide). The leaves are spirally arranged on the shoot and the free part of the leaf which arises from the basal cushion points outwards and upwards. Many examples exist where male cones have been found attached to the foliage, and as a consequence these have never been allocated a name of their own. They are borne singly at the end

---

EXPLANATION OF PLATE 19

Fig. 1. *Elatocladus laxus* (Phillips) Harris; S. 8477; Gristhorpe Plant Bed; × 2.

Fig. 2. *Brachyphyllum mamillare* Lindley and Hutton; S. 1257; Scalby Ness; × 2.

Fig. 3. *Araucarites phillipsii* Carruthers; S. 1343; Gristhorpe Plant Bed; × 1·5.

Figs 4–5. *Brachyphyllum crucis* Harris; Hasty Bank. 4, S. 7049; × 1. 5, S. 3957; male cone (arrowed); × 2.

Figs 6–7. *Elatides williamsonii* (Lindley and Hutton) Nathorst; Gristhorpe Plant Bed; × 2. 6, S. 3009. 7, S. 18067; male cones.

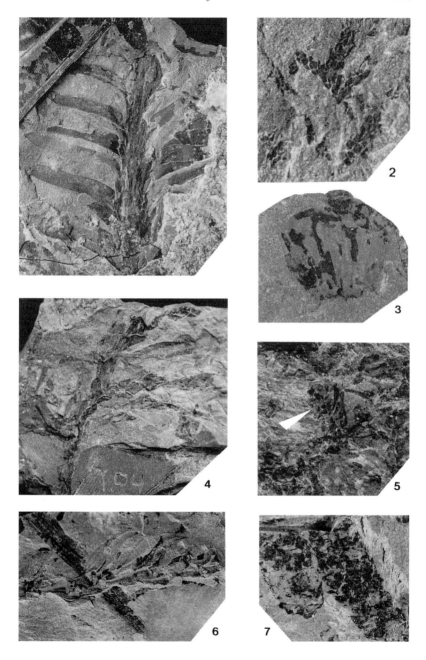

of a leafy shoot, are about 10 mm long and 5 mm wide, and consist of spirally arranged small, rhomboidal microsporophylls. In contrast, only detached female cones have been found and therefore have received the name *Araucarites phillipsii* Carruthers. This applies both to complete cones, which are very rare as they fall apart at maturity, and to detached cone scales (Pl. 19, fig. 3) which are rather more common, particularly in the Gristhorpe Plant Bed. Cone scales are wedge-shaped, with their widest part just below the sharply pointed apex, and are on average 15 mm long and 13 mm wide, with smaller and larger scales also occurring. In the middle of a scale the place of the single seed can normally be observed as a depression. In a normal-sized scale the seeds were typically 10 mm long and 5 mm wide, but in larger or smaller scales their size varied proportionally. The female cones closely resemble those of the living family Araucariaceae, and therefore *B. mamillare* can be attributed to that family without any doubt.

Brachyphyllum crucis Harris (Pl. 19, fig. 4) is a very common fossil at Hasty Bank. Branched leafy shoots can be easily found in many parts of the plant bed. The width of the shoots varies between 5 mm in larger, and 1–3 mm in smaller shoots. The leaves are spirally arranged and may vary considerably in size. The basal cushions in the larger leaves are up to 4·5 mm by 3 mm, and the free part of the leaf is 2–3 mm long. In smaller shoots cushions can be as small as 1·5 mm by 1·0 mm, and the free leaf part only 0·5 mm. However, in all cases the total length of the cushion plus the free part of the leaf is typically twice as long as the leaves are broad.

All male cones (Pl. 19, fig. 5) have been found attached to foliage and thus, have not received a separate name. The cones are about the same size and shape as those of *B. mamillare* Lindley and Hutton and the pollen is typical of the *Classopollis*-type, which allows the cones to be attributed to the extinct Mesozoic family Cheirolepidiaceae (van Konijnenburg-van Cittert 1971). Study of the female cones by Hill (1974*a*) (*Hirmeriella* sp.) supported this supposition, proving affinity with the Cheirolepidiaceae. As both male and female cones are very rare, they will not be described here in any more detail.

Pagiophyllum
Based on leaf morphology and cuticle structure, it is possible that *Pagiophyllum insigne* Kendall (Text-fig. 41A) may also belong to the Cheirolepidiaceae. However, due to the current lack of fructifications this cannot be proven. *P. insigne* is a rare component of the Gristhorpe Plant Bed. The shape and arrangement of the leaves is basically the same as in *Brachyphyllum crucis* Harris, but the free parts of the leaves are much longer. These are typically 12 mm long and 6 mm wide in the middle part of the leaves, 4 mm wide at the base, and tapering apically to a very sharp apex. The substance of the leaves is quite thick.

Elatides

Conifers belonging to this fossil genus (known from the Jurassic and Cretaceous) are considered to belong to the living family Taxodiaceae, based on male and female fructifications. In Yorkshire two *Elatides* species occur, which are similar in many respects but differ in size. *E. williamsonii* (Lindley and Hutton) Nathorst occurs in the Gristhorpe Plant Bed, whilst *E. thomasii* Harris can be found at Hasty Bank.

E. williamsonii (Lindley and Hutton) Nathorst (Pl. 19, figs 6–7) is one of the most common fossils in the Gristhorpe Plant Bed. The leaves are spirally arranged; those at the base or apex of a shoot are rather small, but otherwise the leaves are 6–12 mm long and falcate. The section of the leaf is rhomboidal to nearly square in shape. Usually a keel is visible on the outer side of the leaf. Both male and female cones can be found attached. Male cones (Pl. 19, fig. 7) are borne in groups of three to five at the end of leafy shoots. Immature cones, typically 15 mm long and 5–7 mm wide, expand lengthways at maturity to up to 30 mm, while the width remains

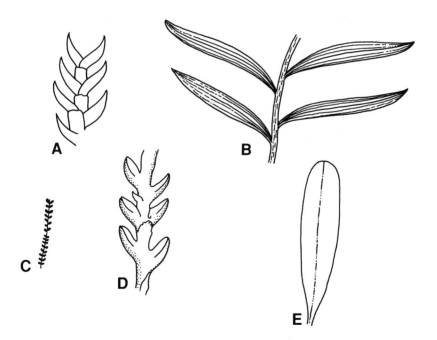

TEXT-FIG. 41. Conifer foliage, after Harris (1979). A, *Pagiophyllum insigne* Kendall; × 0·7. B, *Lindleycladus lanceolatus* (Lindley and Hutton) Harris; × 0·7. C–D, *Geinitzia rigida* (Phillips) Harris. C, shoot; × 0·7. D, detail of a shoot; × 3. E, *Bilsdalea dura* Harris; × 4.

the same. It is understood that after shedding their pollen, the male cones fell off the tree. The cone scales are spirally arranged and consist of a slender stalk and a rhomboidal head, about 1·5 mm in diameter. Female cones (Pl. 20, figs 1, 4) are large and borne singly at the end of a leafy shoot. At maturity the cone has an oval shape, is 40–60 mm long and 20–25 mm wide. The cone scales are spirally arranged, with stalks and rhomboidal heads, about 5 mm wide and 4–5 mm high. They bear five seeds each, but it is unusual to find these attached to the cone scales.

Elatides thomasii Harris (Pl. 20, fig. 2) is a common constituent of the Hasty Bank flora. In general terms the species resembles *E. williamsonii*, but it is smaller in all respects, i.e. leaves, male and female cones. The leaves are up to 5 mm long and 1 mm wide, tapering to an acute apex, are strongly falcate, and almost square in section but not showing a keel. Leaves at both ends of shoots are shorter, and the leaves of small shoots shorter and narrower. Male cones are usually found detached. When mature, the cones are 15–17 mm long and 3–4 mm wide; immature cones are approximately 10 mm long with the same width as at maturity. Cone scales show the same arrangement as *E. williamsonii*, but the rhomboidal heads are only about 1 mm in diameter. Female cones (Pl. 20, fig. 3) are borne terminally on leafy shoots with small, short leaves. The cones are round or oval, 13–20 mm long and 10–15 mm wide. The scales are arranged in the same way as in *E. williamsonii*; but in this case the rhomboidal heads are around 4 mm wide and only 2 mm high. The number of seeds per cone scale is unknown.

Cyparissidium
C. blackii (Harris) Harris (Pl. 20, fig. 5) is a fairly common fossil at Scalby Ness. Usually only unbranched leafy shoots are found, although branched shoots also occur. The leaves are spirally arranged and densely packed, overlapping the bases of the leaves above and normally closely appressed to the stem. The largest leaves known are only 5 mm long and 1 mm wide, whilst smaller leaves are 2 mm by 0·7 mm. Fructifications of *C. blackii* are rarely encountered at Scalby Ness. *Pityanthus scalbiensis*

EXPLANATION OF PLATE 20

Figs 1, 4. *Elatides williamsonii* (Lindley and Hutton) Nathorst; Gristhorpe Plant Bed; × 1. 1, S. 1179; part of female cone attached to a shoot; on the right, is a fertile fragment of *Coniopteris murrayana*. 4, S. 1312; complete female cone.

Figs 2–3. *Elatides thomasii* Harris; Hasty Bank. 2, S. 1429; × 1. 3, S. 1436; female cone (between arrows); × 2

Fig. 5. *Cyparissidium blackii* (Harris) Harris and fragments of *Czekanowskia blackii* Harris; S.1522; Scalby Wyke; × 1

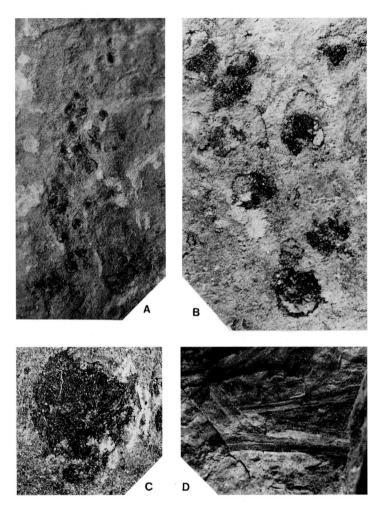

TEXT-FIG. 42. A–B, *Scarburgia hillii* Harris; S. 8481. A, a cone with seeds; × 1. B, close-up of A, showing several cone scales with seeds; × 5. C, *Pityanthus scalbiensis* van Konijnenburg-van Cittert; S. 2966; × 5. D, *Pityocladus scarburgensis* Harris; S. 2324; × 2. All from Scalby Ness.

van Konijnenburg-van Cittert, the male cone (Text-fig. 42C), is of the general coniferous type. It has never been found attached but occurs in close association with the leafy shoots. Mature cones are 8–10 mm long and 3–5 mm wide. The microsporophylls are spirally arranged, consist of a stalk and a head, the latter being around 0·7 mm high and 0·5 mm wide. The female fructification *Scarburgia hillii* Harris (Text-fig. 42A–B) is very rare. It is an elongated, seed-bearing cone (up to 50 mm long and 10 mm wide) with a slender axis bearing fertile appendages. The appendages are made up of a short stalk that expands into a pointed triangular scale with a single rounded seed on its upper surface. At maturity, seed-stones are around 2 mm in diameter. The male and female cones make the attribution of *C. blackii* to the living family Podocarpaceae certain.

Pityocladus
Pityocladus scarburgensis Harris (Text-fig. 42D) is also known from the Scalby Ness locality, where it is locally common. From an old locality along the coast, north of Scarborough (between Scarborough and White Nab; more precise details are not known), good material has been collected which includes long shoots bearing short shoots with scale leaves and needle-like foliage leaves. Female cone scales were also found at that locality at the beginning of this century. Both the needle-like foliage leaves and the cone-scales indicate an affinity to the living family Pinaceae. Today, only foliage leaves can be found at Scalby Ness. The length of the needles is usually greater than 50 mm, and the width about 1 mm in both the middle and apical regions. Near the apex, which is obtusely pointed, the lamina narrows only slightly. It tapers gradually towards the base which is 0·3–0·6 mm wide. The needles were probably stiff in life, probably almost round near the base but becoming flatter above where the midrib is visible.

Lindleycladus
Lindleycladus lanceolatus (Lindley and Hutton) Harris (Text-fig. 41B) is a rare fossil at Hayburn Wyke. It is the only conifer in the Yorkshire Jurassic whose leaves are not scale- or needle-like with a single vein, but instead are, as the specific name indicates, lanceolate with many parallel veins. Female cones belonging to this type of leaf have been found in various parts of the world, but so far, not in Yorkshire. Those female cones place this type of foliage in the fossil family Podozamitaceae, but without fructifications we can only provisionally assign *Lindleycladus lanceolatus* to this family. The leaves of *Lindleycladus lanceolatus* are borne helically on ultimate, caducous, unbranched shoots of strictly limited growth. The leaves are lanceolate, sessile, with a contracted base. The distal end tapers towards a blunt apex, and the veins are numerous, forking near the leaf base only, parallel for most of the lamina but converging towards the apex.

Geinitzia
Geintizia rigida (Phillips) Harris (Text-fig. 41C–D) is another rare fossil at Hayburn Wyke. Its affinities are unknown. The shoot system of this

conifer branches irregularly; all shoots are covered with spirally arranged, small, stiff leaves. The leaf base cushions are normally quite short (1–2 mm long, up to 5 mm on main stems), and the free parts of the leaves commonly 3 mm long and 0·6 mm wide (up to 4·5 mm by 1·2 mm on main stems). The width of the leaves remains uniform almost to the apex, where the lamina is contracted to a mucronate apex.

Bilsdalea

Bilsdalea dura Harris (Text-fig. 41D) is locally common in the Gristhorpe Plant Bed, and can sometimes be found at Hasty Bank. Although its female cones are known, they are so unusual that its affinities remain questionable. Ultimate shoots bearing spirally arranged needle-like leaves are extremely rare, leaves are usually found detached. They are straight, 20–60 mm long, 1·5–2 mm wide and are at once recognized by their glossy surface. The apical part of the leaf tapers only slightly and ends in a rounded, rarely acute apex. The basal part tapers gradually to a base of about one-third of the leaf's full width.

15. PALAEOECOLOGY OF THE RAVENSCAR GROUP

Two alternative approaches have been employed in attempting to identify the palaeoecological controls on plant growth during the Mid Jurassic in Yorkshire. One is concerned with classification of the plant assemblages by type (i.e. in terms of composition, degree of fragmentation, inferred depositional environment), whilst the other investigates the temporal distribution of the flora. The following section is intended as a review of these separate approaches. Field examples are provided where possible, and the four plant beds discussed in this guide are referred to where relevant.

Classification of the plant beds
The first attempt to classify the Yorkshire plant beds in a sedimentological context was carried out by Black (1929) who, based chiefly on plant beds exposed in the cliffs at Scalby and Burniston (just north of Scarborough), concluded that there are two main kinds of plant bed in Yorkshire.

1. 'Drifted' (allochthonous) plant beds. These assemblages are distinctly sorted and dominated by robust, thick-cuticled leaves such as ginkgophytes. The lithology of the beds is typically coarse-grained and associated with distributary channels or sheet sandstones, such as the Scalby Wyke Plant Bed.
2. '*In situ*' (autochthonous) plant beds. These beds are poorly sorted and as a consequence both robust and delicate leaves occur together. These assemblages are typically found in fine-grained sediments deposited in lakes and swamps in the flood-plain environment, such as in the Gristhorpe Plant Bed.

Harris (1952a) refined Black's basic classification using information from all Yorkshire plant beds known to him at the time; he divided the plant assemblages into four main types: (1) those with excellent (that is, well preserved) leaves, (2) coal seams, (3) root beds and (4) dispersed fragments. Harris (1952a, 1953) considered the beds within a depositional context, allowing him to arrange the accumulations into the five types listed below to which have been added observations made by Hill (1974a) and this study. In Harris' scheme, the term *in situ* is restricted to beds whose plants are strictly in the position of growth; this point was stressed by Harris (1952a) because, although there are examples of exceptionally well-preserved plant assemblages in Yorkshire (e.g. the Gristhorpe Plant Bed), a truly autochthonous origin cannot be proven in the absence of roots and stems.

Harris (1952*a*) recognized five types of of plant bed.

1. Beds where plants are preserved in position of growth; upright stems and roots are usually all that remain of the original plant. For example, the '*Equisetum* Bed' at the foot of Beast Cliff, north of Hayburn Wyke, and 5·5 m above base of Sycarham Member, Iron Scar.
2. Fine grained sediments deposited in freshwater lagoons and abandoned channels; fructifications found with leaves and stems, such as the Hasty Bank section and Gristhorpe Plant Bed. Note that Black (1929) ascribed these beds to his *in situ* category.
3. Distributary channels infilled by fine sand, containing a flora similar to that of type 2, albeit with slightly more water-abraded material. Exemplified by the Whitby Long Bight Plant Bed (NZ 905 114).
4. Channel deposits of fine to coarse sand, containing a drifted flora. Water-worn material is prevalent and fragments of charcoal common. Intergrades with type 3. Examples are the Millepore Bed and the plant beds in the Scalby Formation at Yon's Nab and Scalby Wyke.
5. Redeposited plant beds consisting almost entirely of material with thick cuticles, such as the Bilsdale Tripsdale Todd Intake Plant Bed and Westerdale Stockdale Beck waterfall bed.

Discussion

Black's (1929) concept of a two-fold classification for the plant beds of Yorkshire is based upon the assumption that plant fossils behave like sedimentary particles; i.e. because drifted floras are more fragmented and sorted than *in situ* ones, it is assumed that they have been transported further before deposition and therefore represent a more inland source. If this is true, taking Black's idea a step further, it is clear from field observations that drifted floras are mainly associated with coarse-grained sediments, and *in situ* floras with fine-grained deposits. Hill (1974*a*) rationalized this correspondence by assuming that the Yorkshire Jurassic delta behaved in the same way as modern systems, in that sediment of a coarser grain-size is deposited in proximity to the inland source where energy is high, and the finer sediment downstream where the river is more tranquil. It then follows that plant fossils contained within the coarser-grained deposits were derived from further inland and had been comminuted beyond recognition by the time they reached the lower parts of the delta-plain, allowing plants growing locally to be preferentially represented (i.e. the *in situ* flora).

While the notion of an inland flora (drifted) versus delta flora (*in situ*) is the conventional interpretation, there is no evidence that drifted floras grew on upland hills alone. Moreover, modern deltas expose several zones of vegetation, indicating that the 'drifted' and '*in situ*' floras could simply represent two zones of the delta vegetation growing relatively close together. Hill (1974*a*) suggested that the *in situ* flora may have grown on low swampy areas of the floodplain, whilst the drifted flora occupied slightly higher, better drained regions. Therefore, whilst the two floras grew in relatively close proximity, the contrast in ecological conditions imposed by variations in water supply was sufficient to create these rather different floras. Hill (1974*a*) inferred that the common occurrence of lowland *in situ* floras in the succession can be explained by their position of growth along swampy margins of laterally migrating distributary channels: only when the delta subsided and the channels migrated across previously stable land did drifted floras become incorporated into the depositional sequence.

A similar scenario was suggested by Harris (1966) to explain the favoured representation of inland floras in some stratigraphical intervals. Within the otherwise barren, coarse-grained, channel sandstones of the Moor Grit Member (Scalby Formation), thin but laterally extensive plant beds occur containing abundant conifers and presumed taxads (first described by Florin in 1958). Harris (1966) suggested that subsidence of the delta and concomitant relative sea-level rise would have drowned lower areas of the delta-plain and its constituent flora, allowing the preservation of plants from higher levels of the delta.

Although these conclusions may be, in broad terms, correct, their validity may be questioned in the light of field observations made during research carried out for this guide. Firstly, Black's (1929) scheme assumed that the floral assemblages are exclusively drifted or *in situ*, which they clearly are not: in fact there are several examples of single plant beds in Yorkshire (e.g. the Gristhorpe Plant Bed) which, when traced laterally, display features of both types of floral assemblage. Although Black (1929) offered the Gristhorpe Plant Bed as an example of an *in situ* plant bed, roots and stems in the position of growth have not been found, suggesting that the remains could have been transported, even if only over a short distance (Harris 1952*a*). However, even where roots are associated with an assemblage, there may still be a drifted component to the flora. Secondly, by the simple nature of Black's classification, a drifted deposit must represent all plant accumulations thought not to be in the position of growth; this is a somewhat vague bracket to describe what in reality is a range of assemblages that have travelled over different distances. In fact, transportation of a plant organ is commonly variable between species and also between sites. For example, while *Equisetum columnare* occurs *in situ* at Hasty Bank, Iron Scar and north of Hayburn Wyke, at other localities (such as Hayburn Wyke and Cloughton Wyke) it has a disarticulated, drifted appearance. From such points, it may therefore be concluded that, although Black's concept of drifting is essentially correct, the term should ideally be applied to the plant bed in question only in varying degrees of severity.

Yet, even the refined classification of Harris (1952*a*), which stipulated that *in situ* beds must display roots and upright stems, and divided drifted assemblages into four common types, does not account for all possibilities. Of course this is not surprising considering the unique character of many assemblages in terms of their taxonomic diversity, species abundance, degree of fragmentation and leaf physiognomy. The effect of a taphonomic bias upon the original vegetation varies from one depositional environment to another as a product of the sedimentary regime, including matrix type, climate, the hydrological properties of a leaf species and its resistance to degradation and microbial attack. For these reasons, appropriate care must be taken when making palaeoclimatic or other environmental inferences from transported floral remains.

An alternative classification scheme
Table 2 illustrates an alternative classification scheme to separate the Yorkshire flora into plant-bed types (Morgans 1997). In contrast with Harris's is classification, which is based chiefly on the content and degree of fragmentation of the flora itself, this summary considers the flora primarily in terms of the sedimentary facies in which it is contained. Furthermore, the scheme does not distinguish purely autochthonous and allochthonous plant accumulations, as assemblages of these types are frequently found together. The lower part of the Sycarham Member at Iron Scar illustrates this point, as *in situ Equisetum* penetrates a palaeosol in which poorly preserved, fragmentary plant remains are found. In either of the previous classification schemes these floras would have been separated into *in situ* and drifted components. This classification therefore does not 'shoe-horn' the floral assemblages into a small number of plant-bed types conceived by the fragmentation of the plant remains, but instead relies foremost on the recognition of the sedimentary facies within which a variety of species and preservation states may be found.

Changes in the distribution of the flora
In Harris's (1952*b*, p. 209) opinion, the flora is more-or-less evenly distributed spatially across the Yorkshire area, and furthermore does not display a fine stratigraphical zonation which could be useful for correlation purposes. He did, however, find a distinct floral character to each of the Deltaic Series subdivisions, which was evident as a change in the relative abundance of individual species through the succession (excluding those rare species whose range and abundance are poorly known). Harris created five floral groups from this work which he believed had zonal significance. With the addition of a sixth group by van Konijnenburg-van Cittert (1971), the groups are listed below.

Facies	Floral preservation	Description	Example	Black in situ	Black drifted	Harris plant bed type
Variable sand bodies:						
(a) channel lag	wood	Fragmentary, highly sorted assemblages, dominated by wood of various sizes. Best preserved in braided channels.	Channel at base of Saltwick Formation, east Whitby.		✓	3
(b) upper part of point bar	robust/delicate/wood	Diverse, well-preserved, unsorted plant remains.	Scalby Ness Plant Bed, Scalby Ness.		✓	3
(c) upper portion of channel body	robust leaves/wood	Finely disseminated plant fragments, upper parts are also rooted. Some *in situ Equisetum*.	Top of channel body within mid-Saltwick Formation, Hayburn Wyke.		✓	3
Crevasse splays	delicate/wood	Robust, fragmentary plant remains.	Mid-Gristhorpe Member, Yon's Nab.		✓	
		Substantial root systems and *in situ Equisetum*.	Sandstone tiers, Gristhorpe Member, Cloughton Wyke.	✓		1
Interbedded crevasse-splays and flood-plain facies	robust/wood	Mix of robust remains in the crevasse splays and more delicate, unsorted plants in the flood-plain deposits.	Hayburn Wyke Plant Bed.		✓	4
Freshwater lake	robust/wood	Mix of well preserved leaves, stems and wood. *In situ Equisetum* and rootlets commonly found.	Base of Saltwick Formation, Whitby area.		✓	2
Ponds on flood-plain & abandoned channels	robust/wood	Plant remains occur as exceptionally preserved pockets within flood-plain deposits. Wood is pyritized.	Gristhorpe, Hasty Bank, *Solenites* & Scalby Wyke Plant Beds.		✓	2
Interchannel deposits	robust/wood	Plant material is often poorly preserved and fragmentary. *In situ* roots and *Equisetum* common.	Lowermost Sycarham Member, Iron Scar.	✓		1
Back-barrier lagoon	robust/wood	Highly abraded flora, dominated by robust taxa and wood.	Ginkgoalean Plant Bed, Cloughton Wyke.		✓	4
Coastal marsh	robust	Charcoalified wood and leaflets common in sandy parts of this facies, while well preserved leaves are common in the silty beds.	Top of Saltwick Formation, Hayburn Wyke.		✓	4
Various open marine facies	wood	Plant material is uncommon, well sorted, abraded and highly fragmentary.	Scarborough Formation, Cloughton Wyke.		✓	4
Sandy shoreface	robust/wood	Heavy rooting at the top of this facies. Some fragmentary plant remains.	Top of Saltwick Formation, Hayburn Wyke; *Nilssonia* Bed.		✓	1/4
Shallow marine sand belt	wood	Plant material is uncommon, well sorted, abraded and highly fragmentary.	Millepore Bed, Cloughton Wyke.		✓	4

Table 2 An alternative classification scheme to describe the occurrence of plant material in the various sedimentary facies of the Ravenscar Group (Morgans 1997).
Key: ⚘ delicate leaves; ✦ robust leaves; ◉ wood; ⟋ denotes fragmentary plant remains.

1. Species spanning the whole succession without a striking change in abundance, for example, *Brachyphyllum mamillare* Lindley and Hutton and *Coniopteris hymenophylloides* (Brongniart) Seward.
2. Species common in the lower formations but lacking in the uppermost Scalby Formation, including *Equisetum columnare* Brongniart.
3. Species such as *Pachypteris papillosa* (Thomas and Bose) Harris which appear to be confined to the Dogger Formation (immediately below the Ravenscar Group) and/or Saltwick Formation.
4. Species common in the Saltwick Formation which are rare or absent in the Cloughton Formation, only to become common again in the Scalby Formation. For example, *Pseudoctenis oleosa* Harris, *Ptilophyllum pectinoides* (Phillips) Halle, *Baiera furcata* (Lindley and Hutton) Braun and *Pachypteris lanceolata* Brongniart.
5. Species common in the Cloughton Formation, but which are practically absent from the underlying and overlying formations. These include *Ptilophyllum pecten* (Phillips) Morris with male *Weltrichia pecten* (Leckenby) Harris and the female *Williamsonia leckenbyi* Nathorst.
6. Species such as *Ginkgo huttonii* (Sternberg) Heer and *Czekanowskia blackii* Harris, characteristic of the Scalby Formation, but rare or absent from other beds.

According to Harris (1952*b*, p. 207) his results indicate a 'fluctuation and return in the flora', as the Saltwick and Scalby formations are characterized by the relative abundance of one species, and the two members of the Cloughton Formation by the relative abundance of another. In terms of the floral groups listed above, groups 1–3 and 6 are typical of a gradually changing flora through time, and groups 4 and 5 imply a fluctuation and subsequent restoration of the flora. Van Konijnenburg-van Cittert (1971) added the sixth category to Harris's five floral groups in light of the large proportion of unique species contained within the Scalby Formation. This tripartite division of the Ravenscar Group flora into assemblages which typify the Saltwick, Cloughton and Scalby formations, concurs with the findings of Fox-Strangways (1892) who published the first outline of the general stratigraphical relations of the Yorkshire flora. Despite exposing a temporal fluctuation in the flora, Harris (1952*b*) concluded that in light of the apparently equal numbers of xeromorphic to delicate leaves throughout the succession, the flora should be treated as a single entity. This conclusion contrasts with the findings of Florin (1958) and Couper (1958), however, who found that plant accumulations were sufficiently distinct in the Scalby Formation to warrant a tentative temporal division of the flora.

GLOSSARY

Acroscopic: The upper side of (or part of) a leaf; the side directed towards the apex (see Text-fig. 19).

Adnate: Fused.

Anadromic venation: Type of venation where the first *secondary vein* is on the *acroscopic* side; or type of branching where the first pinnule is on the *acroscopic* side (see Text-fig. 19).

Aphlebiform: In the form of aphlebia; i.e. a specialized pinnule (usually at the base of a leaf or *pinna*) of a fern with a shape different from that of the other pinnules.

Auricle: Ear-like, enlarged part of a leaf, near the *acroscopic* leaf base.

Axil: The upper angle between a stem and a leaf.

Basiscopic: The lower side of (part of) a leaf; the side directed towards the base (see Text-fig. 19).

Bract: A modified leaf-like structure.

Capsule: A type of fruit consisting usually of two *valves.*

Commissural furrow: A line between the adnate leaves of the horsetails, indicating where they are joined together.

Cone: A gymnosperm *fructification*, often of compact, ovoid or globular shape (e.g. pine cone).

Corona: Uppermost sterile part of a female bennettitalean *fructification.*

Decurrent: Extending downwards at the point of insertion.

Filiform: With a thread-like shape.

Fructification: A reproductive structure.

Fusain: A component of coal characterized by its fibrous structure, sooty black colour, and silky lustre.

Gynoecium: The female parts of a flower.

Indusium: A shield-like covering over the *sorus* in a fern.

Interseminal scale: The sterile *scales* between the *ovules* in a bennettitalean *fructification.*

Katadromic venation: Type of venation where the first *secondary vein* is on the *basiscopic* side; or type of branching where the first pinnule is on the *basiscopic* side (see Text-fig. 19).

Lamina: Flattened blade part of a leaf.

Lanceolate: Much longer than wide, and tapering to a point.

Lateral vein:	Vein branching from the middle vein (or midrib); see also *Secondary vein.*
Microsporophyll:	A leaf- or scale-like organ bearing one or more microsporangia producing microspores in a pteridophyte and pollen in a seed-plant.
Neuropterid venation:	Venation as in the genus *Neuropteris* (see Text-fig. 43).
Notched:	With indentations.
Ovule:	An unfertilized seed.
Palynomorph:	A microscopic, resistant-walled organic body found in palynological maceration residues: includes pollen, spores, etc., which are insoluble in acids.
Pecopterid pinnule:	Pinnule in the shape of that found in the genus *Pecopteris* (see Text-fig. 43).
Peduncle:	A stalk that bears a fructification or flower.
Peltate:	Shield-like, with the stalk in the middle.
Perianth scales:	*Scales* that together form the 'leaflets' of a flower.
Petiole:	A stalk that connects the flattened blade of a leaf to the stem.
Pinna:	The first-order subdivision of a leaf.
Pinnule:	The ultimate foliar segment of a compound leaf.
Rachis:	The main axis of a leaf.
Receptacle:	The enlarged end of a flower stalk that bears the floral organs in gymnosperms; in ferns, the central stalk of a sorus to which all the *sporangia* are attached.

TEXT-FIG. 43. Schematic drawing of neuropterid (A), pecopterid (B) and sphenopterid (C) venation patterns.

Scale:	A modified leaf-like structure in the form of a scale.
Secondary vein:	Vein branching from the middle vein (or midrib); see also *Lateral vein.*
Sorus:	A group of *sporangia.*
Sphenopterid pinnule:	Pinnule in the shape of that found in the genus *Sphenopteris* (see Text-fig. 43).
Sporangium:	A structure in which spores are produced.
Synangium:	A reproductive unit consisting of fused *sporangia.*
Thallus:	A generalized term for the simple plant body of non-vascular plants; a thallus is not differentiated into roots, stems and leaves.
Truncate:	In the shape of a cut-off trunk.
Valve:	A flap-like structure that is part of a larger structure; e.g. a *capsule* consisting of two valves.
Vein:	A vascular bundle in a leaf as it is seen on the surface.

REFERENCES

ALEXANDER, J. 1992. Nature and origin of a laterally extensive alluvial sandstone body in the Middle Jurassic Scalby Formation. *Journal of the Geological Society, London,* **149**, 431–441.

BATE, R. H. 1959. The Yons Nab Beds of the Middle Jurassic of the Yorkshire Coast. *Proceedings of the Yorkshire Geological Society,* **32**, 153–164.

—— 1965. Middle Jurassic Ostracoda from the Grey Limestone Series, Yorkshire. *Bulletin of the British Museum (Natural History), Geology Series,* **11**, 73–134.

—— 1967. Stratigraphy and palaeogeography of the Yorkshire Oolites and their relationships with the Lincolnshire Limestone. *Bulletin of the British Museum (Natural History), Geology Series,* **14**, 111–141.

BLACK, M. 1929. Drifted plant beds of the Upper Estuarine Series of Yorkshire. *Quarterly Journal of the Geological Society, London,* **85**, 389–439.

BOTT, M. H. P., ROBINSON, J. and KOHNSTAMM, M. A. 1978. Granite beneath Market Weighton, East Yorkshire. *Journal of the Geological Society, London,* **135**, 535–543.

CITTERT, J. H. A. van 1966. Palaebotany of the Mesophytic II. New and noteworthy Jurassic ferns from Yorkshire. *Acta Botanica Neerlandica,* **15**, 284–289.

COUPER, R. A. 1958. British Mesozoic microspores and pollen grains. A systematic and stratigraphic study. *Palaeontographica, Abteilung B,* **103**, 75–179.

DONATO, J. A. 1993. A buried granite batholith and the origin of the Sole Pit Basin, UK Southern North Sea. *Journal of the Geological Society, London,* **143**, 255–258.

FISHER, M. J. and HANCOCK, N. J. 1985. The Scalby Formation (Middle Jurassic, Ravenscar Group) of Yorkshire: reassessment of age and depositional environment. *Proceedings of the Yorkshire Geological Society,* **45**, 293–298.

FLORIN, R. 1958. On Jurassic taxads and conifers from north-west Europe and eastern Greenland. *Acta Horti Bergiana,* **17**, 257–402.

FOX-STRANGWAYS, C. 1892. Jurassic rocks of Britain, Vol. 1, Yorkshire. *Memoir of the Geological Survey of the United Kingdom,* ix+551 pp.

GOWLAND, S. and RIDING, J. B. 1991. Stratigraphy, sedimentology and palaeoecology of the fining-upwards cycles and aligned gutter marks in the Middle Lias (Lower Jurassic) of Yorkshire. *Proceedings of the Yorkshire Geological Society,* **48**, 375–392.

GRADSTEIN, F. M., AGTERBERG, F. P., OGG, J. G., HARDENBOL, J., VAN VEEN, P., THIERRY, J. and HUANG, Z. 1995. A Triassic, Jurassic and Cretaceous time scale. 95–126. *In* BERGGREN, W. A., KENT, D. V., AUBRY, M. P. and HARDENBOL, S. (eds). *Geochronology time scales and global stratigraphic correlation.* SEPM Special Publication No. 54.

HANCOCK, N. J. and FISHER, M. J. 1981. Middle Jurassic North Sea deltas with particular reference to Yorkshire. 186–195. *In* ILLING, L. V. and HOBSON, G. D. (eds). *Petroleum geology of the continental shelf of North-West Europe.* Institute of Petroleum, London.

HARRIS, T. M. 1940. *Caytonia. Annals of Botany, London, New Series,* **4**, 713–734.

—— 1941. *Caytonanthus*, the microsporophyll of *Caytonia*. *Annals of Botany, London, New Series*, **5**, 47–58.

—— 1951. The relationships of the Caytoniales. *Phytomorphology*, **1**, 29–39.

—— 1952*a*. The zonation of the Yorkshire Jurassic flora. *The Palaeobotanist*, **1**, 207–211.

—— 1952*b*. Floral succession in the Estuarine Series of Yorkshire. *International Geological Congress, 'Report of the Eighteenth Session Great Britain 1948'*, **10**.

—— 1953. The geology of the Yorkshire Jurassic Flora. *Proceedings of the Yorkshire Geological Society*, **29**, 63–71.

—— 1961. *The Yorkshire Jurassic flora I. Thallophyta-Pteridophyta*. British Museum (Natural History), London, 212 pp.

—— 1964. *The Yorkshire Jurassic flora II. Caytoniales, Cycadales and Pteridosperms*. British Museum (Natural History), London, 191 pp.

—— 1966. Dispersed cuticle. *Palaeobotanist*, **14**, 102–105.

—— 1969. *The Yorkshire Jurassic flora III. Bennettitales*. British Museum (Natural History), London, 186 pp.

—— 1979. *The Yorkshire Jurassic flora V. Coniferales*. British Museum (Natural History), London, 166 pp.

—— 1980. The Yorkshire Jurassic fern *Phlebopteris braunii* (Goeppert) and its reference to *Matonia* R.Br. *Bulletin of the British Museum (Natural History), Geology Series*, **33**, 295–311.

—— 1981. Burnt ferns from the English Wealden. *Proceedings of the Geologists' Association*, **92**, 47–58.

—— MILLINGTON, W. and MILLER, J. 1974. *The Yorkshire Jurassic Flora IV. Ginkgoales and Czekanowskia*. British Museum (Natural History), London, 150 pp.

HEMINGWAY, J. E. 1974. Jurassic. 161–223. *In* RAYNER, D. H. and HEMINGWAY, J. E. (eds). *The geology and mineral resources of Yorkshire*. Yorkshire Geological Society, Leeds.

—— and KNOX, R. W. O'B. 1973. Lithostratigraphic nomenclature of the Middle Jurassic strata of the Yorkshire Basin of north-east England. *Proceedings of the Yorkshire Geological Society*, **39**, 527–535.

—— and RIDDLER, G. P. 1982. Basin inversion in North Yorkshire. *Transactions of the Institute of Mining and Metallurgy, Section B*, **91**, 175–186.

HESSELBO, S. P. and JENKYNS, H. C. 1995. A comparison of the Hettangian to Bajocian successions of Dorset and Yorkshire. 105–150. *In* TAYLOR, P. D. (ed.). *Field geology of the British Jurassic*. Geological Society, London.

HILL, C. R. 1974*a*. Palaeobotanical and sedimentological studies on the lower Bajocian (Middle Jurassic) flora of Yorkshire. Unpublished Ph.D. thesis, University of Leeds.

—— 1974*b*. Further plant fossils from the Hasty Bank locality. *Naturalist, London*, **929**, 55–56.

—— and KONIJNENBURG-van CITTERT, J. H. A. van 1973. Species of plant fossils collected from the Middle Jurassic plant bed at Hasty Bank, Yorkshire. *Naturalist, London*, **925**, 59–63.

HOGG, N. M. 1993. A palynological investigation of the Scalby Formation (Ravenscar Group, Middle Jurassic) and adjacent strata from the Cleveland Basin, north-east Yorkshire. Unpublished Ph.D. thesis, University of Sheffield.

KANTOROWICZ, J. D. 1990. Lateral and vertical variation in pedogenesis and other early diagenetic phenomena, Middle Jurassic Ravenscar Group, Yorkshire. *Proceedings of the Yorkshire Geological Society*, **48**, 61–74.

KENT, P. 1980. *British regional geology: eastern England from the Tees to the Wash.* 2nd edition. H.M.S.O., London.

KIRBY, G. A. and SWALLOW, P. W. 1987. Tectonism and sedimentation in the Flamborough Head region of north-east England. *Proceedings of the Yorkshire Geological Society*, **46**, 301–309.

KNOX, R. W. O'B. 1973. The Eller Beck Formation (Bajocian) of the Ravenscar Group of NE Yorkshire. *Geological Magazine*, **110**, 511–534.

KONIJNENBURG-van CITTERT, J. H. A. van 1971. In situ gymnosperm pollen from the Middle Jurassic of Yorkshire. *Acta Botanica Neerlandica*, **20**, 1–96.

—— 1989. Dicksoniaceous spores in situ from the Jurassic of Yorkshire, England. *Review of Palaeobotany and Palynology*, **61**, 273–301.

—— 1996. Two *Osmundopsis* species from the Middle Jurassic of Yorkshire and their sterile foliage. *Palaeontology*, **39**, 719–731.

LEEDER, M. R. and NAMI, M. 1979. Sedimentary models for the non-marine Scalby Formation (Middle Jurassic) and evidence for late Bajocian/Bathonian uplift of the Yorkshire Basin. *Proceedings of the Yorkshire Geological Society*, **42**, 461–482.

LIVERA, S. E. and LEEDER, M. R. 1981. The Middle Jurassic Ravenscar Group ('Deltaic Series') of Yorkshire: recent sedimentological studies as demonstrated during a Field Meeting, 2–3 May 1980. *Proceedings of the Geologists' Association*, **92**, 241–250.

LOVIS, J. D. 1975. *Aspidistes thomasii*—a Jurassic member of the Thelypteridaceae. *Fern Gazette*, **11**, 137–140.

MILSOM, J. and RAWSON, P. F. 1989. The Peak Trough—a major control on the geology of the North Yorkshire coast. *Geological Magazine*, **126**, 699–705.

MORGANS, H. S. 1997. Early to Middle Jurassic vegetation, climate change and stratigraphic development in north-eastern Europe. Unpublished D.Phil. thesis, University of Oxford.

MUIR, M. D. 1964. The palaeoecology of the small spores of the Middle Jurassic of Yorkshire. Unpublished Ph.D. thesis, University of London.

NAMI, M. 1976. An exhumed Jurassic meander belt from Yorkshire, England. *Geological Magazine*, **113**, 47–52.

PARSONS, C. F. 1977. A stratigraphic revision of the Scarborough Formation (Middle Jurassic) of north-east Yorkshire. *Proceedings of the Yorkshire Geological Society*, **41**, 203–222.

—— 1980. Aalenian and Bajocian correlation chart. *In* COPE, J. C. W., DUFF, K. L., PARSONS, C. F., TORRENS, H. S., WIMBLEDON, W. A. and WRIGHT, J. K. (eds). A correlation of the Jurassic rocks of the British Isles. Part Two: Middle and Upper Jurassic. *Geological Society, London, Special Report*, **15**, 3–20.

POWELL, J. H. and RATHBONE, P. A. 1983. The relationship of the Eller Beck Formation and the supposed Blowgill Member (Middle Jurassic) of the Yorkshire basin. *Proceedings of the Yorkshire Geological Society*, **44**, 365–373.

RAWSON, P. F. and WRIGHT, J. K. 1992. *The Yorkshire coast*. Geologists' Association, Field Guide No. 34, London, 117 pp.

RIDING, J. B. and WRIGHT, J. K. 1989. Palynostratigraphy of the Scalby Formation

(Middle Jurassic) of the Cleveland Basin, north-east Yorkshire. *Proceedings of the Yorkshire Geological Society*, **47**, 349–354.

SELLWOOD, B. W. and JENKYNS, H. C. 1975. Basins and swells and the evolution of an eperic sea (Pliensbachian–Bajocian of Great Britain). *Journal of the Geological Society, London*, **131**, 373–388.

SPICER, R. A. and HILL, C. R. 1979. Principal components analysis and correspondence analyses of quantitative data from a Jurassic plant bed. *Review of Palaeobotany and Palynology*, **28**, 273–299.

SYLVESTER-BRADLEY, P. C. 1953. A stratigraphical guide to the fossil localities of the Scarborough District. 19–43. *In*: *The natural history of the Scarborough district, 1*. Scarborough.

TAYLOR, T. N. and TAYLOR, E. L. 1993. *The biology and evolution of fossil plants*. Prentice-Hall, Eaglewood Cliffs, NJ, 981 pp.

THOMAS, H. H. 1925. The Caytoniales, a new group of angiospermous plants from the Jurassic rocks of Yorkshire. *Philosophical Transactions of the Royal Society of London, Series B*, **213**, 299–363.

WRIGHT, J. K. 1977. The Cornbrash Formation (Callovian) in North Yorkshire and Cleveland. *Proceedings of the Yorkshire Geological Society*, **41**, 325–346.

YOUNG, G. and BIRD, J. 1822. *A geological survey of the Yorkshire coast*. Clark, Whitby, iv+332 pp.

ZHOU ZHIYAN 1991. Phylogeny and evolutionary trends of Mesozoic ginkgoaleans, a preliminary assessment. *Review of Palaeobotany and Palynology*, **68**, 203–216.